HIGH EFFICIENCY DEEP GRINDING

High Efficiency Deep Grinding

TECHNOLOGY, PROCESS PLANNING, AND ECONOMIC APPLICATION

Dr-Ing. Taghi Tawakoli

English translation edited by C. G. Barrett

DeBeers Industrial Diamond Division

VDI-Verlag GmbH

Mechanical Engineering Publications Limited
LONDON

First Published (in German) 1990
© VDI-Verlag GmbH, Düsseldorf, Germany

English language edition 1993
© T. Tawakoli

ISBN 0 85298 820 6

A CIP catalogue record for this book is available from the British Library.

Typeset by Santype International Ltd, Salisbury, Wilts
Printed and bound in Great Britain by
Biddles Ltd, Guildford and King's Lynn

Foreword

High-efficiency deep grinding does not conform to the findings of conventional grinding. The results of theoretical and practical research carried out to date show that increasing individual parameters such as wheel speed, depth of cut, or specific removal rate brings about an increase in grinding temperature which can cause damage to the workpiece. These parameters are set at a very high level in high-efficiency deep grinding, but, in spite of this, the grinding temperatures are lower than in conventional grinding.

This shows that previous theoretical and practical findings from conventional grinding cannot be applied equally to high-efficiency deep grinding. Machine and tool requirements in high-efficiency deep grinding are similarly different from those which apply in conventional grinding.

This book describes and discusses the technological requirements and theoretical principles of high-efficiency deep grinding. The role of machine settings and their influence on the process is examined and explained. Against this backdrop, guidance is given for the practical application of the process.

The techniques and findings presented here are based on experience gained in a wide range of industrial and research projects. The publication itself stems from a dissertation accepted by Bremen University. For his support in this work I wish to express my special thanks to Prof. Dr-Ing. G. Werner, the former head of the Manufacturing Technology Section of the Production Engineering Department.

In December 1992 this work was revised and supplemented in the light of further research. I gratefully acknowledge the support and assistance of De Beers Industrial Diamond Division in the publication of this volume.

Taghi Tawakoli
Bremen, January 1993

Contents

Symbols and abbreviations

A		Proportionality factor
A_k	mm^2	Contact area
a_e	mm	Depth of cut
a_{ed}	mm	Dressing infeed
a	m^2/s	Temperature conductivity
B		Boron nitride
B_b		Bond displacement factor
b_k	mm	Contact layer width
b_s	mm	Wheel width
C_1	mm^3	Static edge density relative to unit depth $z = 1$
C_k	mm^{-3}	Number of grains per unit volume
C_{hl}	Ws/mm^3	High-efficiency deep grinding factor
d_k	mm	Grit diameter
d_s	mm	Wheel diameter
d_{se}	mm	Equivalent wheel diameter
E'	J/mm^3	Specific energy
E_a		Start of engagement
E_e		End of engagement
F_{kmax}	N	Grain pull-out force
F'_n	N/mm	Specific total normal force
F'_t	N/mm	Specific total tangential force
F''_{ges}	N/mm^2	Total grinding force relative to unit area
F''_n	N/mm^2	Normal force relative to unit area
F''_t	N/mm^2	Tangential force relative to unit area
f_1		Chip-space reduction factor
G	mm^3/mm^3	Grinding ratio
GE		Up-cutting
GL		Down-cutting
HEDG		High-efficiency deep grinding
h_{cu}	μm	Chip thickness
h_{eq}	μm	Equivalent chip thickness
K	%	Grit concentration relative to volume
K_1, K_2		Proportionality factor
K_e, K_n		Proportionality factor

K_r		Heat conversion factor
K_s		Edge shape factor
K_z		Grit protrusion factor
L		Dimensionless contact length index
l_k	mm	Contact length
l_s	mm	Grinding length
N_{mom}		Instantaneous number of cutting edges
N_{stat}	mm^{-2}	Static number of cutting edges per unit area
n, n_s	min^{-1}	Wheel rotational speed
n_k		Number of grits on wheel periphery
n_{kl}		Number of grits on 1 mm of wheel width
n_{sw}		Swing number
p		Exponent of function $S_{stat}(z)$
P_c	kW	Spindle output
P_{ges}	kW	Total output
P_L	kW	No-load output
P_{br}	kW	Braking power
P'_c	kW/mm^3	Specific grinding power
Q'_{sb}	mm^3/(mm · s)	Specific sharpening removal rate
Q	W	Heat flow into workpiece
Q'_w	mm^3/(mm · s)	Specific removal rate
Q'_{wd}	mm^3/(mm · s)	Specific dressing volume
Q'_{w151}	mm^3/(mm · s)	Specific removal rate for B151 grit
q		Exponent of function $N_{stat}(z)$
q_m		Elongation factor of CBN grits
q_w	W/mm	Quantity of heat flowing into workpiece per unit of time and length
R_a	μm	Arithmetical average peak-to valley height
R_{as}	μm	Arithmetical average peak-to-valley height of the grinding wheel
R_p	μm	Smoothing depth
R_z	μm	Average peak-to-valley roughness
R_{zs}	μm	Average peak-to-alley roughness of the grinding wheel
r_a	mm	Outside radius
r_i	mm	Inside radius
r_r	mm	Roll radius
r_s	mm	Grinding wheel radius
r_t	mm	Swing radius

T_1		Temperature measuring point 1
t_d	s	Dressing time
t_{sch}	s	Time to remove the contact layer
V_{bz}	mm^3	Protrusion volume for one grit
V_{cu}	mm^3	Chip volume
V_k	mm^3	Grit volume
V_{ges}	mm^3	Total volume of half a grit, including the corresponding chip space
V'_{sch}	mm^3/mm	Specific contact layer volume
V_{sp}	mm^3	Volume of chip-space for a grit with a protrusion $z = d_k/2$
V_{spl}	mm^3/mm	Specific chip space volume
V_{spb}	mm^3	Chip space not available if grit protrusion $z < d_k/2$
$V_{sp, ges}$	mm^3	Total volume of chip space on periphery of grinding wheel
V_{spv}	mm^3	Chip volume on periphery of grinding wheel
V'_{spv}	mm^3/(mm $\cdot$ s)	Maximum removal rate
V_{spz}	mm^3	Grit volume and corresponding chip space volume for a grit with protrusion $z < d_k/2$
v_c	m/s	Wheel speed
v_{cd}	m/s	Wheel speed when dressing
v_R	m/s	Dressing roll speed
v_t	mm/s	Speed of swing of the dressing roll
v_w	mm/s	Workpiece speed
w_m	μm or mm	Average mesh size
z	mm	Grit protrusion
z_1	mm	Difference in grit protrusion
z_{krit}	mm	Critical grit protrusion
α		Exponential coefficient
α_e	degrees	Engagement angle
α_{max}	degrees	Maximum angle of swing
β_k		Proportionality factor
β_u	degrees	Edge engagement angle in up-grinding
γ	degrees	Rake angle
γ	s^{-1}	Deformation speed
ε		Grindability coefficient
ε_3		Exponential coefficient

θ	°C	Surface zone temperature
θ_0	°C	Initial temperature
θ_z	°C	Temperature below the surface zone
λ	W/(m · K)	Thermal conductivity
μ		Cutting force ratio
v		Poisson's ratio
ξ		Degree of filling of chip-space
ρ	kg/m^3	Density
σ_{max}	N/mm^2	Maximum tangential stress
ω	s^{-1}	Angular velocity

1

Introduction

Because of its special characteristics, namely great accuracy and the high surface quality it can generate in the machining of hardened materials, grinding has hitherto been used as a finishing operation. Thanks to modern machines and extremely hard cutting tool materials such as PCBN (polycrystalline cubic boron nitride) and PCD (polycrystalline diamond), the same results can now often be achieved with other machining methods, such as turning, milling, shaving, and so on. When comparable component qualities can be obtained by different techniques, the choice of machining method will be determined by economic criteria.

The development of grinding technology in recent years bears out the accuracy of a study carried out some years ago in the USA, which forecast the future trend in this field of engineering (1)*. It predicted that crucial developments would take place by the year 2000 in this particular area of production engineering. High-efficiency deep grinding with CBN tools confirms that this is, indeed, the case. The technology, which has been developed from a combination of the high-speed and creep-feed grinding techniques, can be characterised by its ability to achieve high stock removal rates and produce a high-quality finish on the workpiece.

Compared with conventional reciprocating grinding, creep-feed grinding is characterised by improved surface quality, lower temperature effects on the surface layer of the workpiece, higher compressive residual stresses, more favourable dynamic behaviour of the component, and reduced tool wear (2)–(7). In general, creep-feed grinding is an economic method for machining profiles, which can usually be produced in a single operation. Creep-feed grinding is performed at low wheel speeds, such as are customary with reciprocating grinding.

Much research has been devoted to determining the effect of higher wheel speeds on the course of the grinding process and on the results produced. The reduction in grinding forces and surface roughness are the most important benefits obtained by increasing the wheel speed (8)–(11). Disadvantages of increased wheel speeds include higher surface temperatures (in the speed range investigated of $v_c < 80$ m/s) and the greater

---

* References are listed together at the end of the book (see p. 131).

risks arising from the fracturing of the tool. Recent developments in grinding machines and tools are, however, characterised by a trend towards higher wheel speeds.

In retrospect, the use of creep-feed grinding at high wheel speeds seems to be a logical development towards improving the efficiency of the grinding process. The combination of the two techniques enables material removal rates to be achieved which are more than a hundred times greater than those attainable with conventional reciprocating grinding.

The main features of high-efficiency deep grinding are: large depths of cut, high feed rates (with the resulting high specific stock removal rates), and high wheel speeds. The enhanced performance associated with this method has only been made possible by the availability of suitably designed grinding machines and CBN grinding tools capable of operating at wheel speeds (v_c) in excess of 125 m/s.

However, high-efficiency deep grinding results not only in increased productivity due to reduced manufacturing time, but also in an improvement in the process in terms of tool wear, specific energy requirements, and the surface quality of the workpiece. These benefits are conditional on the use of grinding machines, tools, control systems, and auxiliary equipment (coolant, wheel-cleaning, balancing, and dressing systems) appropriate to the high-efficiency technique. Despite its fundamental advantages, high-efficiency grinding has, to date, not yet achieved wide application in practice. This may be explained by the general lack of information available regarding the design, monitoring, and control aspects of this relatively new technology.

The purpose of this book is to remedy this lack of knowledge, and to describe the mechanical and technological requirements of high-efficiency deep grinding, using plunge-cut surface grinding as an example. With regard to high stock removal rates, the effect of grinding and dressing parameters on the process variables, and on the results obtained, is demonstrated by practical examples. In addition, the user is provided with recommendations on how the method should be applied, and the machine tool manufacturer is provided with suggestions for improved machine and tool design.

2

The state of the art

2.1 CREEP-FEED GRINDING AS AN INTRODUCTION TO HIGH-EFFICIENCY SURFACE GRINDING

As the creep-feed grinding technique has developed, and its areas of application widened, conventional machining methods, such as form milling and turning, can now be replaced to some extent (6)(7)(12).

Following the introduction of the process into industrial manufacturing, it took a long time for creep-feed grinding to win general acceptance. Larger contact lengths, higher total energy conversion, and higher grinding forces led to the widely held view that creep-feed grinding might cause thermal damage to the surface of the workpiece and the surface zone (2)–(3).

Scientific research succeeded in throwing light on the technological mechanisms, thus explaining the advantages of creep-feed grinding compared with conventional reciprocating grinding. In particular it was the important finding that, in creep-feed grinding, during the removal process the newly produced surface is exposed to a substantially lower thermal load than in reciprocating grinding, which led to the widespread adoption of the creep-feed grinding method (4)–(7).

Research into creep-feed grinding at higher stock removal rates could only be carried out with the introduction of new machines and grinding tools and the higher peripheral speeds possible with such tools. This research paved the way to the current use of high-performance machines with modern CNC control systems for carrying out semi-automated machining operations.

The further development of creep-feed grinding at greater stock removal rates gave rise to the term 'high-efficiency deep grinding'. The machines used for high-efficiency deep grinding have a high spindle output and are fitted with a stepless speed control. At wheel speeds in excess of 100 m/s special grinding wheels must be used. Very few aluminium oxide wheels can be used at peripheral speeds as high as 125 m/s. The development of new abrasives, such as CBN (cubic boron nitride), has resulted in significant advances in the field of modern high-efficiency grinding. Very high wheel speeds ($v_c > 100$ m/s) are possible when CBN abrasives are used in specially made grinding wheels (for example, with a metal bond).

In order to achieve high stock removal rates, the machine settings used in high-efficiency deep grinding are much higher than in conventional grinding. The increase in the depth of cut and the specific stock removal rate results, for example, in an increase in temperature in the contact zone between workpiece and tool (**8**)–(**10**)(**13**). However, a high contact-zone temperature should by no means be equated with a high temperature on the newly produced workpiece surface. Ignorance of this fact is the reason why many potential users shy away from the use of high-efficiency deep grinding. The effects of increased wheel peripheral speed have been known for more than 50 years (**11**): lower grinding forces and reduced tool wear, together with higher surface quality, all of which are due to smaller chip cross-sections (**14**)(**15**). Furthermore, as the peripheral speed of the grinding wheel rises, there is an increase in the temperature in the contact zone between the wheel and the workpiece.

Figure 2.1 shows the workpiece surface-zone temperature as a function of wheel speed, v_c. Excessive temperatures in this zone are undesirable, because of the damage to the workpiece which is likely to occur. If increasing the wheel speed leads to elevated temperatures in both the

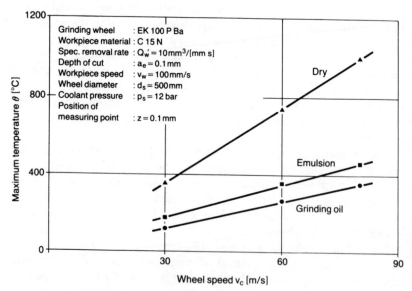

Fig. 2.1   Workpiece surface-zone temperature as a function of wheel peripheral speed (8)

contact and the surface zone, as has been observed in many studies (8)–(10)(13)(14), the wheel speed may be increased only to the threshold value at which no damage is caused to the workpiece.

Apart from its effects on contact length, mean chip thickness, number of kinematic cutting-edges, grinding forces, and surface quality, depth of cut a_e also affects the temperature in the contact zone (4)–(10)(13)–(15). This is due to the fact that, at a constant feed rate and constant wheel speed, the specific removal rate Q'_w increases with depth of cut. This is accompanied by greater energy conversion, leading to higher temperatures in the contact zone. If the specific removal rate is kept constant by correspondingly reducing the workpiece speed as the depth of cut increases, the surface-zone temperature rises and reaches its maximum at a given depth of cut. If the depth of cut is increased further, the temperature decreases once more (4)–(6).

Figure 2.2 shows the curve for the maximum temperature on the newly produced surface as a function of the depth of cut at a constant specific removal rate. The area in the left of the figure is characteristic of reciprocating grinding. The workpiece surface-zone temperature rises in response to the higher temperatures resulting from an increasing depth of cut, as well as in response to a prolonged reaction time caused by the lower workpiece speed (4)–(7). It is known from conventional grinding practice that an increase in the depth of cut is accompanied by an increase in temperature. In the case of creep-feed grinding, this should mean a continuing rise in the workpiece surface temperature at larger depths of cut and lower workpiece speeds. Under these conditions, temperatures of several thousand degrees Celsius might occur (Fig. 2.2, point A). However, investigations have shown that from a given depth of cut, and with the specific removal rate held constant ($Q'_w = a_e \times v_w$), the tendency for the surface temperature to increase is reversed. Consequently, in the creep-feed grinding range (the right-hand portion of the figure), the workpiece surface temperature falls as the depth of cut is increased.

The reasons for the rising temperature curve in reciprocating grinding, and the decreasing temperature curve in creep-feed grinding, lie in the conditions of contact between the grinding wheel and the workpiece. With the two grinding processes in question, these are fundamentally different. In creep-feed grinding the contact length is 20–60 mm. The corresponding workpiece speeds are less than 10 mm/s. In reciprocating grinding the contact length is 0.2–2 mm and the workpiece speed is 50–500 mm/s.

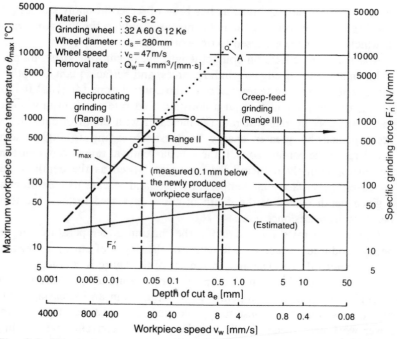

Fig. 2.2 Workpiece surface-zone temperature as a function of depth of cut and proportionally decreasing workpiece speed (4)

This means that the thermal reaction time in creep-feed grinding is 250–1000 times longer than in reciprocating grinding. At the same time, however, the heat flow per unit of surface area is greatly reduced, so that, although in creep-feed grinding a greater total quantity of heat flows into the unit area of workpiece surface, this takes place over a much longer period. The result is a lower surface-zone temperature (4).

Increasing the speed of the workpiece, under otherwise constant machining parameters, reduces the temperature in the newly produced workpiece surface (9)(11)(13)(15). Figure 2.3 shows the effect of workpiece speed and specific removal rate on the surface temperature in cylindrical grinding (9). The temperature was measured with sheathed thermocouples during the grinding of ball-bearing steel 100 Cr 6, very close to the newly produced workpiece surface. If Q'_w is kept constant, an increase in v_w causes the temperature to fall. Increasing the peripheral speed of the

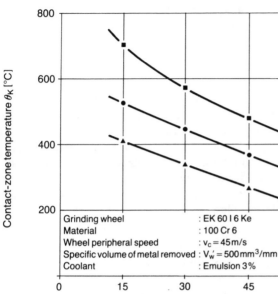

Fig. 2.3 **Decreasing contact-zone temperature with increasing workpiece speed for various specific removal rates** (9)

workpiece from 15 to 60 m/min produces a temperature reduction of approximately 200–300°C, depending on the value of Q'_w. With an increasing specific removal rate, or with a greater depth of cut, and a constant workpiece peripheral speed, the temperature increases. Similar results have been obtained in surface grinding (11)(13).

The benefits of the low workpiece surface temperature at high workpiece speeds cannot be fully utilised in creep-feed grinding, as the process requires that the workpiece speed be kept low. Only high-efficiency deep grinding enables full use to be made of the advantage of a low workpiece temperature at a high workpiece speed.

Gühring was the first to propose theoretical arguments as to the processes occurring in high-efficiency deep grinding (11). The development and production of high-efficiency/high-speed grinding machines enabled these theoretical considerations to be translated into practice (13).

2.2 TECHNOLOGICAL REQUIREMENTS FOR COMBINING CREEP-FEED GRINDING WITH HIGH-SPEED GRINDING

The previously held view with regard to creep-feed grinding, that the greater depth of cut would have a negative effect on the surface zone of the workpiece, was reflected in the assumption that the high cutting speed in high-speed grinding would have a similar effect. This is one of the reasons why there has been some hesitation in combining the two techniques and putting them to practical use.

With his force and temperature model, Werner (16) demonstrated analytically the possibility of combining creep-feed grinding and high-speed grinding to create the high-efficiency deep grinding process. According to this model, the thermal and mechanical loads at the surface zone of the workpiece are found to be as follows.

(a) *Thermal loading of the workpiece surface*

At large depths of cut ($a_e \geqslant 0.8$ mm) and relatively low workpiece speeds ($v_w \leqslant 20$ mm/s), the maximum temperature at a point close to the newly produced surface of the workpiece can be calculated from the following formula

$$T_{max} = (> a) = \frac{K_T}{\varepsilon} \times (C_1)^\gamma \times (v_c)^{2-2\varepsilon} \times (Q'_w)^{2\varepsilon - 1 - \alpha}$$
$$\times (a_e)^{1/2 - \varepsilon - \alpha/2} \times (d_s)^{1/2 - \varepsilon + \alpha/2} \tag{2.1}$$

Assuming good thermal grindability, the following values may be assigned to the three exponential coefficients occurring in this expression

$\varepsilon = 0.9$ This analytically-defined coefficient takes into consideration the ratio of deformation energy to frictional energy in chip formation in terms of heat generation. When the frictional energy is low, the value of ε approximates to 1.0, while, with a high proportion of frictional energy, the values of ε lie close to 0.5.

$\gamma = 0.1$ This coefficient, also analytically defined, reflects the influence on heat generation of the distribution of cutting edges (number and shape). When thermal grindability is favourable, i.e., where the proportion of frictional energy is low, the values of γ approximate to 0. With poor grindability, i.e., a high proportion of frictional energy, the values of γ lie between 0.5 and 1.0, depending on the shape of the cutting edges.

$\alpha = 0.6$–1.0 This empirically-determined coefficient allows for the quantity of energy discharged from the contact zone in the form of heat with the chips. It depends on the thermomechanical properties of the material, and is relatively great when the frictional energy is low. It also depends on the process parameters a_e, v_w, and d_s.

This gives us the following quantitative expression for the temperature of the workpiece surface during high-efficiency deep grinding

$$T_{max} = \frac{K_T}{0.9} \times (C_1)^{0.1} \times (v_c)^{0.2} \times (Q'_w)^{+0.2 \text{ to } -0.2}$$

$$\times (a_e)^{-0.7 \text{ to } -0.9} \times (d_s)^{+0.1 \text{ to } -0.1} \tag{2.2}$$

From this it can be concluded that

(1) With increasing depth of cut a_e and proportionally reduced workpiece speed v_w (the values of C_1, v_c, Q'_w, and d_s being constant), the temperature of the workpiece surface drops significantly.
(2) With increasing specific removal rate Q'_w, produced by a higher workpiece speed v_w (the values of C_1, v_c, a_e, and d_s being constant), the temperature may rise slightly or fall slightly, depending on the efficiency with which heat is dissipated by the chips.
(3) With increasing wheel speed v_c (the other parameters being constant), there is only a very slight increase in temperature.

Taking into consideration the measures for increasing Q'_w and v_c, an optimum combination of the two parameters is feasible (assuming good grindability). This conclusion was reached by Werner (17), on the assumption that, as the wheel speed rose, there would be a steady increase in the surface temperature of the workpiece.

More recent findings, set out in the present work, show that the surface temperature of the workpiece does not, in fact, rise steadily as the wheel speed increases. On the contrary, after reaching a certain value, any further increase in wheel speed is accompanied by a reduction in the surface temperature of the workpiece.

(b)   *Mechanical loading of the workpiece surface*

According to Werner (**16**), the relationship of the specific total grinding force per unit of grinding wheel width can be represented as follows

$$F' = \frac{K_n}{\varepsilon} \times (C_1)^\gamma \times \left(\frac{Q'_w}{v_c}\right)^{2\varepsilon - 1} \times (a_e \times d_s)^{1-\varepsilon} \tag{2.3}$$

With the same exponential coefficients, $\varepsilon = 0.9$ and $\gamma = 0.1$ (good thermal grindability), the following quantitative equation can be derived for the specific total grinding force in high-efficiency deep grinding

$$F' = \frac{K_n}{0.9} \times (C_1)^{0.1} \times \left(\frac{Q'_w}{v_c}\right)^{0.8} \times (a_e \times d_s)^{0.1} \tag{2.4}$$

The following conclusions can be made with regard to high-efficiency deep grinding.

(1)   With increasing depth of cut a_e and proportionally reduced workpiece speed v_w, there is a slight increase in total grinding force.
(2)   With increasing specific removal rate Q'_w, due to a higher workpiece speed v_w, the cutting force increases.
(3)   With increasing wheel speed v_c, there is a marked decrease in cutting force.

The downward trend in the total cutting force with increasing v_c is as marked as the increase in total cutting force with increasing Q'_w. This means that simultaneously raising the specific removal rate and the wheel speed is advantageous with regard to a low specific total grinding force.

The relationships described above demonstrate that, in high-efficiency deep grinding, the thermal and mechanical loads on the workpiece surface remain limited. It must, therefore, be possible to combine creep-feed grinding and high-speed grinding (**17**).

2.3   DISCUSSION OF VARIOUS HIGH-EFFICIENCY SURFACE GRINDING PROCESSES

The following grinding techniques may be regarded as forerunners of high-efficiency deep grinding:

−   creep-feed grinding with continuous dressing (CD grinding);
−   high-speed grinding with CBN wheels;
−   high-efficiency belt-grinding.

These grinding techniques are, therefore, considered in somewhat greater detail below.

2.3.1 Creep-feed grinding with continuous dressing (CD grinding)

Grinding with wheels which are dressed continuously during the machining operation (i.e., with Continuous Dressing, abbreviated to CD grinding) is a high-efficiency grinding technique which has already established itself in practice.

Figure 2.4 shows the principle of CD grinding. The amount removed during dressing, i.e., the reduction in the diameter of the wheel, is continuously compensated by the machine by means of a suitable infeed device. This enables flat surfaces and profiles to be machined accurately. In this process, sharp cutting edges are always available for the removal of stock. The quantity of material removed by dressing must always be greater than the wear on the grinding wheel. CD grinding is generally used on materials which are difficult to machine. A typical example is the grinding of turbine blades made of high-temperature-resistant materials (17)(18). The main advantage of the CD method is the great reduction in

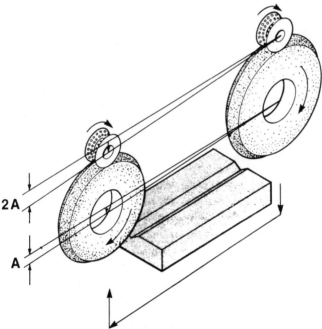

Fig. 2.4 The principle of CD grinding (18)

production time which can be achieved on account of the higher specific removal rate (**19**).

The constant dressing of the wheel in CD grinding ensures that the tool remains true to shape. This can greatly reduce geometrical errors. A further advantage of this process is its reduced power consumption. Figure 2.5 compares the specific energy requirements for grinding with and without continuous dressing (**20**).

The greater specific energy requirement in conventional grinding is attributable to the blunting of the grains during the cutting process. The blunted, flattened grains have larger surfaces and result in increased friction during grinding. This, in turn, gives rise to a higher temperature in the contact zone.

The CD technique is used mainly in the creep-feed grinding of profiled components. The most commonly used dressing tool is the diamond roll (**2**)(**4**). The dresser infeed is usually 0.2–2 μm per revolution of the grinding wheel (**18**). The optimum value for the dresser infeed depends on the

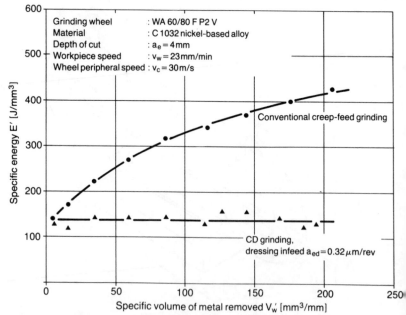

Fig. 2.5   Comparison of specific energy requirements in CD grinding and conventional grinding (according to (20))

surface quality required, the workpiece material, the grinding wheel speci-fication, the depth of cut, and the specific removal rate. Higher removal rates are achieved by increasing the dresser infeed, which produces a rougher and therefore 'sharper' wheel topography. However, it then has to be accepted that the finish on the workpiece will also be rougher.

The relatively high surface roughness values compared with those achieved by grinding without continuous dressing are, however, not gov-erned solely by the greater roughness of the grinding wheel. Continuous dressing causes grains to be loosened in the bond, and these are dis-lodged as soon as contact is made. These loose grains leave deep traces of scoring in the workpiece. Depending on the requirements of the given task, a compromise has to be struck between a high removal rate and a rough finish on the workpiece. Specific removal rates Q'_w of 50–70 mm^3/(mm · s) can be achieved in CD grinding. These are higher by a factor of ten than those achieved in conventional creep-feed grind-ing, and, other conditions being equal, in many cases exceed the removal rates attainable with CBN wheels.

Notwithstanding the wide acceptance of CD grinding in industry, in most cases the technique is inferior to high-efficiency deep grinding, which can achieve a much higher specific removal rate. Moreover, the frequent wheel-changes necessary in CD grinding are a disadvantage as they result in lost productivity and demand a labour input.

2.3.2 High-speed surface grinding with CBN wheels

Cubic boron nitride (CBN) is a newly developed grit. It is produced by synthesis in a high-pressure, high-temperature process (over 50 kbar and around 2000°C) (21). Since their introduction to the market in 1969, there has been a great increase in the use of CBN tools, initially for tool grinding, and now for production grinding as well (22).

CBN grinding tools are characterised by a range of properties (23)–(27)

- *High wear-resistance.* The hardest material after diamond, CBN wears much more slowly than aluminium oxide or silicon carbide. This means that the accuracy-to-shape of the tool is maintained even after long periods of use.

- *Thermal stability.* CBN has a high degree of stability up to tem-peratures of about 1000°C, and is therefore suitable for machining steel. Diamond, on the other hand, can be used only up to a temperature of 800°C.

– *Chemical stability.*   Despite higher temperatures and pressures in the contact zone, CBN grit is less prone to chemical reactions, especially if grinding oil is used as coolant (**26**).

– *Good thermal conductivity.*   CBN possesses excellent thermal conductivity. It is, in fact, three times higher than that of copper (**27**). This property leads to the rapid dissipation of heat from the contact zone between the tool and the workpiece.

– *Cool cutting.*   When CBN grit is used, the chip-formation process differs from that with conventional abrasives, as this material is, in general, characterised by sharper cutting edges with a rake angle γ of -60 to -70 degrees (with conventional grits, $\gamma = -85$ to -90 degrees). This means that 'hot', easily flowing chips are not formed, but, rather, discontinuous 'cold' chips. This means that steel grinding with CBN abrasive is characterised by relatively low temperatures and higher engagement forces for the individual grains (**12**).

The high performance of CBN grinding tools can be attributed to two properties in particular.

In the first place, the high wear-resistance of CBN results in a significant reduction in downtime, especially that due to wheel changing and dressing. The low wear also has a beneficial effect on trueness-to-shape and geometrical accuracy. In the second place, the different construction of the wheel compared with aluminium oxide wheels, in that it consists of a metal hub with a layer of CBN abrasive, allows higher wheel speeds to be used. A high wheel speed is one of the main requirements for achieving high specific removal rates.

There are two disadvantages, however. Firstly, the relatively high cost of CBN must be mentioned; secondly, the extremely hard CBN grit is difficult and costly to dress. The dressing process is generally divided into two operations: profiling and sharpening. The geometrical accuracy of the tool (profile and roundness) is first assured by a profiling stage and, secondly, a sharpening operation is carried out to lower the level of the bond and create sufficient chip space.

Single-layer electroplated CBN wheels are generally used without dressing. They are simpler to manufacture, and, because of their steel core, wheels of this kind can be used at wheel speeds of up to 200 m/s and more.

CBN wheels with a vitrified bond are simpler to dress than metal or resin bond wheels. The pore spaces in the abrasive coating facilitate the splitting and dislodging of the grains from the bond. Subsequent sharpening is not always necessary in the case of CBN wheels with a vitrified bond.

In spite of high costs, and the not-yet-completely-resolved problems of dressing, the use of CBN wheels is economic in many applications. Two examples will illustrate the industrial use of CBN wheels with a high removal rate.

The first example relates to the high-speed grinding of collet chucks using electroplated CBN wheels, *cf.* Figure 2.6 (**28**). The 1 mm wide and up to 11 mm deep slot is machined into HSS material using an electroplated CBN wheel running at a wheel speed of 135 m/s and a feed rate of 2 m/min, i.e., a maximum specific removal rate Q'_w of 250–350 mm^3/(mm · s) is achieved.

Figure 2.7 shows a second example involving grinding grooves in HSS drill bits. Resin-bond CBN and aluminium oxide wheels were used to machine the hardened blank. The figure also provides comparative data for the grinding ratio G and the average roughness, as well as the relative production costs and the maximum specific removal rate achieved for each groove. In each case, the CBN wheel produces the more favourable results.

High-speed grinding with CBN wheels is understood to mean primarily a high, but not precisely defined, wheel speed. In the literature, wheel speeds of 63 m/s, 100 m/s, and even up to 200 m/s are described as high-speed grinding (**13**)(**30**). When high-speed grinding also achieves high removal rates, the process becomes high-efficiency deep grinding. In practice the term 'high-speed grinding' is often used to refer to grinding processes with a high specific removal rate. A high wheel speed is essential for high-efficiency deep grinding, but is not, in itself, sufficient for high specific removal rates.

2.3.3 High-efficiency belt grinding
In recent years belt grinding, too, has been developed into a high-efficiency grinding process. In addition to its widespread use for deburring and grinding complex parts, belt grinding with high specific removal rates, capable to some extent of competing with other cutting processes, has been developed on a growing scale (**31**).

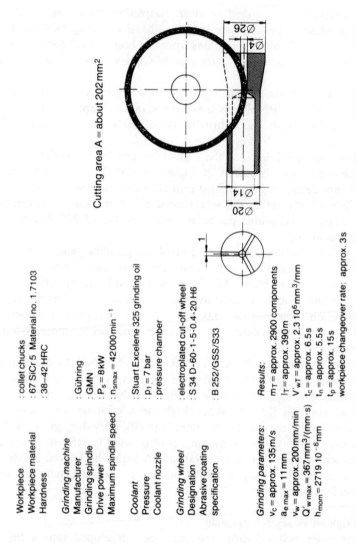

Cutting area A = about 202 mm²

Workpiece   : collet chucks
Workpiece material   : 67 SiCr 5 Material no. 1.7103
Hardness   : 38–42 HRC

Grinding machine
Manufacturer   : Gühring
Grinding spindle   : GMN
Drive power   : P_s = 8 kW
Maximum spindle speed   : n_{smax} = 42000 min^{-1}

Coolant
Pressure   : Stuart Excelene 325 grinding oil
Coolant nozzle   : p_1 = 7 bar
                             : pressure chamber

Grinding wheel
Designation   : electroplated cut-off wheel
Abrasive coating   : S 34 D-60-1-5-0.4-20 H6
specification   : B 252/GSS/S33

Grinding parameters: *Results:*
v_c = approx. 135 m/s m_T = approx. 2900 components
$a_{e\,max}$ = 11 mm l_T = approx. 390 m
v_w = approx. 200 mm/min V'_{wT} = approx. 2.3 10^6 mm³/mm
$Q'_{w\,max}$ = 367 mm³/(mm·s) t_c = approx. 6.5 s
h_{mom} = 2719 10^{-6} mm t_n = approx. 5.5 s
                                        t_p = approx. 15 s
                                        workpiece changeover rate: approx. 3 s

Fig. 2.6   High speed grinding of collet chucks (28)

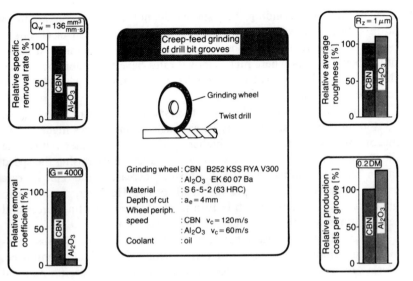

Fig. 2.7   High speed grinding of drill-bit grooves with CBN and aluminium oxide grinding wheels (29)

In belt grinding, the tool takes the form of an endless, flexible belt of fabric coated with abrasive, which is brought into contact with the surface to be machined. At the workpiece surface, the belt is supported by contact elements, such as a roll or shoe. High-performance machines are now built for high-efficiency belt grinding, capable of supplying an output of about 1 kW for every millimetre of belt width (for speeds v_c of up to 60 m/s) (29)(31).

As a result of new developments and constant improvements to the abrasive materials, flexible substrates, and bonding agents, belt grinding is now used as a high-performance manufacturing technique in a wide variety of applications.

Due to their great hardness and toughness, zirconium aluminium oxide and the new 3M grit are particularly suitable for belt grinding. Polyester substrates enable high belt speeds of up to about 60 m/s to be used. New adhesives have contributed to improvements and greater strength at the joints in the belt, which were a frequent cause of premature rupture. Because of its very high thermal stability, phenolic-resin-based adhesive has proved to be an effective bonding agent (29).

In normal belt grinding, belt speeds do not exceed a maximum of 30 m/s. In high-efficiency belt grinding, a maximum speed of 60 m/s is possible. These relatively modest speeds are due to the limited strength of the belt. In cases where high removal rates have to be obtained nonetheless, relatively large cutting forces have to be applied, which give rise to increased grit failure and wear.

In the dry grinding of steel, the specific removal rate Q'_w is 70–100 mm^3/(mm · s). This may be as much as three times higher when machining cast iron. However, higher specific removal rates reduce the life of the tool. The specific removal rate is increased by raising the speed of the belt. At a limited belt speed, a further increase in the removal rate can be achieved by increasing the depth of cut. However, this leads to greater contact lengths, which in turn result in greater cutting forces and temperatures, as well as considerable belt wear.

Belt grinding is generally performed without a coolant, because conventional grinding belts soon lose their strength if a coolant is employed. Steels are now machined without a coolant, at high specific removal rates and suffering no thermal damage **(31)(32)**.

The benefits and capabilities of belt grinding are offset by the following disadvantages.

- At present, the method can be used only for deburring and surface grinding, i.e., it is unsuitable for profile grinding.
- Even though the belts are relatively long, they must be changed frequently.
- The rough surface finish it produces limits the applications for belt grinding, especially where there are stringent requirements in terms of surface quality.
- As the wear on the belt increases, blunted and rounded grains give rise to high frictional forces which can cause thermal damage to the workpiece.

Belt grinding does not, therefore, compete directly with high-efficiency deep grinding. It is not able to produce high quality finishes or shapes.

3

Task and objective

The combination of high workpiece quality and a high removal rate is one of the main objectives and a deciding factor in most chip-forming machining processes. High-efficiency deep grinding, which combines creep-feed grinding and high-speed grinding, meets the requirement for high removal rates and high workpiece quality. During the last two decades, much work has been carried out to improve grinding machines, control systems, and grinding wheels (particularly with regard to abrasives and bonds), and to enhance our understanding of the fundamental technological aspects of high-efficiency deep grinding. The results of this work have been considerable, but as yet there has been no comprehensive treatment of the various phases of development, or an understandable and adequate description of the technological principles of the process, the machines required for high-efficiency deep grinding, and the relevant limiting criteria. The purpose of the present work is to make good this lack of information about high-efficiency deep grinding.

The main emphasis will be laid on the following crucial aspects of the grinding process.

(a) The requirements which have to be met by the grinding machine, tools, and auxiliary systems.
(b) Determination of the limiting criteria for high-efficiency deep grinding and the development of an analytical model for calculating the maximum possible specific removal rate.
(c) The relationship between process variables, workpiece characteristics, and grinding parameters.

The general requirements for high-efficiency deep grinding, as these relate to the grinding machine, the power required, speeds, and the grinding wheel, will be discussed first. Attention will then be turned to further requirements and conditions associated with the auxiliary systems (coolant apparatus, coolant supply and cleaning systems, and profiling and sharpening devices), which have been designed and built for the process. The technological basis for achieving high removal rates will also be discussed by way of theoretical considerations.

A theoretical and analytical approach is adopted for determining the limiting criteria applicable to high-efficiency deep grinding. As a result of such considerations, an analytical model is developed for establishing the maximum attainable specific removal rate, with the emphasis on electroplated CBN wheels. The theoretically-determined limit values are compared with those obtained in practical investigations, and the accuracy of the model is examined.

Against this background, the significant process variables, measured quantities and results, and their relationship to the grinding parameters, are discussed and tested in practical conditions. In examining the various relationships and influencing factors, emphasis is placed on spindle output, grinding forces, surface roughness, temperature, residual stresses, grit size, and grinding direction.

Finally, this information is used to provide a clear guide for the practical application of the process.

4

Technological principles and requirements for high-efficiency surface grinding

The present state of knowledge makes it clear that, under certain technological conditions, high removal rates can be combined advantageously with high wheel speeds. These conditions, which in the past have only been approached from the empirical standpoint, can also be determined analytically on the basis of modern grinding technology.

4.1 DEFINITION OF THE PROCESS AND MACHINE SETTINGS

From a technical viewpoint, surface grinding can be divided into three distinct techniques, differentiated in terms of the process employed and the maximum removal rate:

- reciprocating grinding;
- creep-feed grinding;
- High-Efficiency Deep Grinding (HEDG).

HEDG is understood to mean creep-feed grinding at high wheel speeds and high removal rates. A common feature of creep-feed grinding and HEDG is a large depth of cut ($a_e = 0.2$–25.0 mm or more). The essential characteristics distinguishing HEDG from creep-feed grinding are the significantly higher wheel speed (**12**)(**33**)–(**36**)(**38**) and the much higher workpiece speed employed in HEDG. Table 4.1 compares the most important machine settings employed in the different processes. High-efficiency deep grinding is characterised by the following features:

- high wheel speed ($v_c > 80$ m/s);
- large depth of cut ($a_e = 0.2$–25.0 mm or more);
- medium to high workpiece speed ($v_w = 0.5$–10.0 m/min or more).

In addition to the machine settings, intensive application of coolant is also required (oil, high pressure, high throughput, and a separate cleaning system).

Table 4.1 Comparison of the machine settings and specific removal rates characteristic of different surface grinding methods

Machine settings	Method		
	Reciprocating grinding	Creep-feed grinding	HEDG
Depth of cut a_e	low 0.001–0.05 mm	high 0.1–30 mm	high 0.1–30 mm
Workpiece speed v_w	high 1–30 m/min	low 0.05–0.5 m/min	high 0.5–10 m/min
Wheel speed v_c	low 20–60 m/s	low 20–60 m/s	high 80–200 m/s
Specific removal rate	low 0.1–10 mm³/(mm · s)	low 0.1–10 mm³/(mm · s)	high 50–2000 mm³/(mm · s)

The specific removal rate is the arithmetical product of the depth of cut and the workpiece speed

$$Q'_w = a_e \times v_w \qquad (4.1)$$

It follows that, if the technical conditions are adhered to, the removal rates attainable with high-efficiency deep grinding exceed by a factor of 100 or more those which can be achieved by conventional reciprocating and creep-feed grinding.

4.2 GRINDING MACHINE REQUIREMENTS

The following features specific to HEDG arise from the conditions characteristic of the process:

- short machining times ($t_s = 0.1$–10.0 s);
- high grinding forces;
- increased dynamic sensitivity, with correspondingly increased balancing problems;
- increased braking effect caused by the coolant on the rotating wheel;
- stricter demands on the dressing process;
- increased safety requirements;

– increased environmental protection requirements when oil is used as coolant.

In order to satisfy these factors, in addition to meeting the technological conditions, the grinding machine used for HEDG also has to satisfy a number of requirements. These comprise:

– high power output;
– suitable spindle mounting and control systems;
– suitable machine control system;
– rigid bed design;
– optimised coolant-supply system;
– integrated balancing system;
– suitable dressing system.

As a starting point for the tests described below, a HEDG machine meeting all these requirements was designed and built as the prototype for a production machine. This machine is described in greater detail in section 5.1 (Fig. 5.1). For certain tasks the spindle output is determined by the following factors:

– the required specific removal rate;
– geometry of the profile to be machined;
– wheel speed;
– grinding wheel specification;
– coolant and its viscosity;
– properties of the workpiece material.

The geometry of the profile which is to be machined influences the braking effect in the grinding zone. The deeper the wheel penetrates into the workpiece, and the greater the length of contact, the greater is the lateral braking effect on the wheel. The braking effect also depends on the viscosity of the coolant. The spindle output P_c which is required can be roughly calculated by applying the following equation

$$P_c = P_{br} + Q'_w \times b_s \times c_{hl} \tag{4.2}$$

where $Q'_w \geqslant 100 \ mm^3/(mm \cdot s)$.

When an oil coolant is being used, with a profile depth a_e of 6 mm, a wheel diameter d_s of 400 mm, a wheel speed v_c of 140–160 m/s and a medium-size CBN grit as the abrasive, the following values can be inserted for the braking power P_{br} and the cutting performance factor C_{hl}

$P_{br} = 30$ kW

$C_{hl} = 25$ Ws/mm^3

Equation (4.2) may then be written

$$P_c = 30 \text{ kW} + Q'_w \cdot b_s \cdot 0.025 \text{ kWs/mm}^3 \qquad (4.3)$$

where b_s is the width of the grinding wheel in mm, and Q'_w is the specific removal rate in mm^3/(mm · s).

In most cases, about two thirds of the spindle output is consumed in overcoming the braking effect of the (oil) coolant on the wheel rotating at high speed in the groove. Because of the high proportion of the braking power and the high removal rates, the spindle output needed in HEDG is about 3–6 times greater than that required in conventional creep-feed grinding. In general, the spindle output for medium wheel diameters (around 400 mm) should not be less than 50 kW. Smaller wheel diameters result in shorter contact lengths and lower friction, and, therefore, less power is needed.

4.3   TOOL REQUIREMENTS

In high-efficiency deep grinding the tool is subjected to high resultant forces. In the first place, it is exposed to greater centrifugal forces, and secondly, the higher removal rates result in increased contact and grinding forces. The wheel hub and the abrasive layer must be able to withstand these forces.

At very high wheel speeds ($v_c > 120$ m/s), only wheels with a metal hub (steel, aluminium, or aluminium/plastic composites) are used. In these cases, the abrasive coating usually consists of single-layer (electroplated) or multi-layer (metal-bond) CBN systems (Fig. 4.1).

Because of the limited strength of the coating, CBN wheels with a resin or vitrified bond are only used at wheel speeds (v_c) of up to 120 m/s.

For economic reasons efforts are also being made to use conventional wheels with a vitrified or bakelite bond for high-efficiency deep grinding. The wheel speed is normally limited to 80 m/s. However, with a special structure, and subject to specific approval, conventional wheels may also be used up to a maximum of 120 m/s.

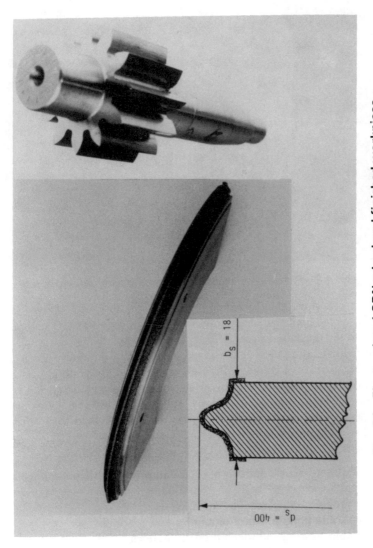

Fig. 4.1 Electroplated CBN wheel and finished workpiece

Fig. 4.2 Partially coated CBN wheel for angular grooves

Higher peripheral wheel speeds are not possible with conventional wheels, as the high centrifugal forces would lead to bursting. The limiting factor is the tangential stress at the inside diameter (**38**)(**39**).

Notwithstanding this limitation, conventional grinding wheels can also be used to advantage in high-efficiency deep grinding, and an increase in the removal rate by a factor of 100 is possible. Resin bond wheels have performed particularly well, as they exhibit a favourable chip-space configuration, and have less tendency to load than vitrified-bond conventional wheels.

Essential requirements for HEDG wheels are listed below:

- low out-of-balance due to wheel-mounting (concentricity < 10 percent of grit size);
- low wobble (< 5 percent of grit size);
- secure and uniform wheel-mounting;
- suitable dressing and touch-dressing systems.

These conditions relate both to the manufacture of the wheels and to their mounting and dressing systems. They are vital to the HEDG process, and adherence to these conditions often governs the success of the process.

Grinding wheels produced by attaching segments with an adhesive are less suitable for HEDG because of the limited strength of the adhesive bond and the interrupted cut. The same applies to slotted wheels. Although these are superior to unslotted wheels insofar as the coolant can be introduced into the contact zone more effectively, they tend to give a dynamic stimulus to vibrations, which can have a harmful effect on the grinding process.

In the case of electroplated, non-cylindrical CBN wheels with steep tapered contours, partial coating of the wheel has proved beneficial (Fig. 4.2), as it has the advantages that the lateral cutting pressure is reduced and the supply of coolant is improved. The conditioning of the wheel is extremely important, i.e., dressing, touch-dressing, and sharpening. The dressing of (conventional) grinding wheels ensures the roundness and profile of the wheel, and the same is achieved by the shaving, or 'touch-dressing', of CBN wheels. Sharpening produces sharp cutting edges evenly distributed over the surface of the tool.

During the grinding tests it proved to be very important to ensure that the requirements relating to the grinding wheel were met.

4.4 COOLANT SUPPLY

In the grinding process, large quantities of energy are converted in the contact zone between the grinding wheel and the workpiece. Some of this acts on the workpiece in the form of heat and may result in damage to the workpiece. In order to reduce this danger, grinding is normally carried out with a coolant. This has five purposes.

(a) *Lubrication*
Lubrication reduces friction and contact forces, thus reducing energy conversion.

(b) *Cooling*
That part of the energy which causes the workpiece to be heated must be dissipated by the coolant by a convection process.

(c) *Chip removal*
The coolant has to remove the chips produced by the grinding operation from the working zone.

(d) *Cleaning the grinding wheel*
     If particles from the workpiece lodge in the surface of the wheel, this
     upsets the grinding process. High-pressure flushing of the grinding
     wheel can prevent loading.

(e) *Cooling the spindle, motor, bearings, etc.*
     The high performance and high speed of the spindle, motor, and any
     bearings mean that there must be a constant supply of coolant. To
     prevent the need for additional cooling units, the coolant circulates
     in a closed circuit attached to the machine [42].

According to DIN 51385, coolants are classified as non-water-mis-
cible, water-miscible, and water-mixed (**43**)(**44**). Grinding oil is the non-
water-miscible coolant generally adopted for grinding. Of the other two
groups, water-miscible emulsion is the most commonly used coolant.

Oil reduces grinding forces, especially at high removal rates, and also
produces an improved surface finish (**43**)(**44**). Along with the advantages
of oil, there are, however, environmental and tolerance problems (**45**),
which have to be combatted by such measures as completely enclosing
the machine, extraction, filtration, and disposal of the oil mist. The fact
remains that, for high-efficiency deep grinding, oil is more suitable than
emulsion. The lubricating action of the oil reduces the grinding forces,
which are relatively high owing to the high removal rates. In addition,
the use of an oil coolant reduces bond erosion, thereby preventing the
grit from being dislodged prematurely (**44**).

Another reason why an effective lubricating action is particularly
important in high-efficiency deep grinding is the need to reduce the
energy and temperature level in the contact zone. This reduces the
importance of the cooling effect, as heat which is not generated in the
first place does not need to be dissipated. In HEDG, grinding oil has
proved to be a suitable coolant, which has to be supplied at high pressure and in sufficient quantity.

The evacuated chips are filtered out of the oil in a treatment unit. The
oil must be delivered to the grinding zone by suitable pumps, pipes, and
nozzles. In HEDG, this calls for a special coolant-supply system (see
Fig. 4.3). This system, designed both for production purposes and
for the tests, provides for a total throughput of 350 litres of grinding
oil per minute at a pressure of 14 bar. This ensures that grinding wheels
up to a width of about 25 mm can receive a coolant flow of at least

Fig. 4.3   Coolant supply system developed for high-efficiency deep grinding

14 l/(mm · min). This is ten times more than that used in conventional reciprocating and creep-feed grinding.

For the separate wheel-flushing process, the throughput in this system is 150 l/min at a pressure of 20 bar. After comparing gravity filters, centrifuging, and suction-band filters, the latter system was chosen as it was not only technically efficient but also cost-effective and of sufficient capacity.

Both the cooling and the cleaning of the wheel require costly systems which are described in greater detail below.

4.4.1   Suitable supply systems

An appropriate and efficient supply of coolant to the grinding zone is one of the essential requirements for successful high-efficiency deep grinding.

In reciprocating grinding, which is characterised by lower removal rates and shorter lengths of contact, the supply of coolant presents less of a problem than in creep-feed grinding and HEDG. In HEDG, the shorter grinding times, greater lengths of contact, and higher wheel speeds accentuate the need for the coolant to be supplied correctly.

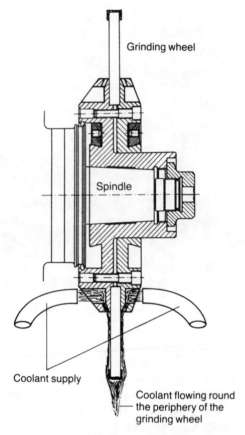

Grinding wheel

Spindle

Coolant supply

Coolant flowing round
the periphery of the
grinding wheel

Fig. 4.4(a)   System for delivering coolant to thin grinding wheels

In conventional grinding, the speed at which the coolant is delivered
should approximate to the peripheral speed of the wheel (**46**)(**47**). In these
circumstances, the tangentially-flowing coolant adheres to the surface of
the wheel which then carries it into the contact zone. This coolant transmission system continues to act advantageously even when the speed of
the coolant flow $v_{kss} = 0.6 \times v_c$ (**47**).

When the delivery speed exceeds 30–40 m/s, it becomes difficult to
control the coolant flow in such a way as to suit the process. The jet
tends to become turbulent (**40**), and the supply process has to be

increased to uneconomically high levels. A different supply system is, therefore, preferable for high-efficiency deep grinding. The delivery of copious supplies of coolant from a number of nozzles floods the grinding zone in the area at which the wheel enters the workpiece. The wheel, therefore, passes through a layer of coolant before reaching the workpiece, and its rough surface carries the adhering coolant into the contact zone.

Whatever the method of coolant supply, it is very important that the nozzle system is designed in a way which is appropriate to the process. Besides free-jet nozzles there are pressure-chamber nozzles, shoe nozzles, and coolant delivery through the grinding wheel. This latter technique has been used for internal cylindrical grinding with vitrified bond wheels (**48**). Free delivery of the coolant via the side of the rotating wheel is also possible. In high-performance/high-speed grinding, shoe nozzles have proved particularly effective, but in HEDG it is the flooding system with a number of nozzles which has provided the best results.

With thin grinding wheels, it is possible, by making use of the centrifugal forces, to achieve better coolant delivery at lower pressure and a reduced application rate. A special flange or auxiliary wheels have to be designed for this purpose (Fig. 4.4(a)). Depending on the peripheral speed of the grinding wheel, the supply hoses and nozzles are positioned such that the centrifugal forces cause the coolant to run down the sides of the wheel and spread round its periphery in, or just before, the grinding throat. If the grinding wheel is greater than 2 mm in thickness, this method does not adequately cool the wheel's peripheral face, and for this reason this delivery technique can only be used for thin wheels. The main advantage of this system is that it does not require any shaped parts or baffles to be positioned with great accuracy relative to the wheel profile.

As an example for optimum coolant delivery and cooling action, Fig. 4.4(b) shows a wing cell rotor with six angular-ground grooves machined from the blank in one pass to a depth of 55 mm (**42**). Feed speed for the workpiece was 300 mm/min. The feed chosen was relatively small as spindle output (50 kW) was at its maximum. If greater spindle power were available, the feed could be greater. There was no evidence of any thermal effect on the surface zone of the workpiece and the grinding results achieved were excellent.

In brief, there is no shape of nozzle which suits every application. A tailor-made solution has to be found and tested on each occasion. The following points must be borne in mind.

Fig. 4.4(b)   Grooves ground in a wing cell rotor

- Whenever possible, the coolant should be delivered to all sides of the wheel at a pressure of about 10 bar.
- In general, the nozzle system should be fixed relative to the grinding wheel.
- In every application a guide must be mounted in front of the workpiece to ensure optimum delivery of the coolant into the contact zone from the very start of the grinding operation.

4.4.2   Wheel cleaning

The effective removal of chips lodged in the chip-spaces is often an essential feature in grinding, as the chip-forming process is hampered if the chip-spaces are loaded. Further consequences are higher grinding forces, increased power consumption, a rougher finish, and increased wheel wear. All these factors are not only detrimental to the process and its economics, but are also likely to have an adverse effect on the results achieved (**49**). Special tests carried out by Sperler have established the connection between wheel loading and the occurrence of burning on the surface of the workpiece (**50**).

In HEDG, which is performed with shorter cycle times and mainly with the use of more expensive tools, loading of the wheel can have a far more harmful effect than in conventional grinding. Because the grinding times are so short, the machining cycle cannot be interrupted in the middle to enable loading to be remedied by a measure such as dressing.

The tendency to load depends on almost all the grinding variables, including the workpiece material. Carbon steels with a low carbon content and chrome steels, for example, have a marked tendency to loading (**49**).

No detailed investigations have yet been carried out into the relevance of the problems to HEDG. However, a number of conclusions can be drawn from the results available so far.

- Vitrified bond wheels have a greater tendency to load than resin bond wheels. Tests carried out by Lauer-Schmaltz produced the same results (**49**).
- Single-layer electroplated CBN wheels show less tendency to load than wheels with multiple layers of abrasive.
- Vitrified bond CBN wheels have pore voids with a more complicated shape than those of resin bond or electroplated wheels. The relatively deep and angular pore voids promote the mechanical adhesion and retention of the chips, thus aggravating the problem of loading.
- On grinding wheels with no pore voids, the chips themselves create chip-spaces of optimum shape, which exhibit little tendency to load.
- Unlike emulsion, an oil coolant does not cause much loading.
- The presence of solid lubricants (e.g., metal soaps) in the bond reduces both loading and grinding forces (**51**)(**52**).

Suitable cleaning nozzles supplying coolant at an appropriate pressure are an effective means of preventing the chip-spaces from becoming loaded. By continuous flushing of the wheel at 35 bar, Grabner was able to double the specific removal rate (**44**). In the case of HEDG, the optimum results have been obtained with the cleaning system developed for this technique supplying coolant at an average pressure of 15–20 bar. The apparatus shown by way of example in Fig. 4.5 is for cleaning the sides of a thin grinding wheel. The system was used to clean a form wheel with a low angle of inclination, used to machine 25 mm deep grooves. The same system was also successfully applied to the peripheral face of the wheel.

The apparatus shown in Fig. 4.5 consists of two opposite arrays of 25 nozzles each, at 1 mm intervals. The symmetrical disposition of the

Fig. 4.5 System of nozzles for cleaning the sides of grinding wheels

nozzles prevents any axial deformation of the thin grinding wheel, as well as any excitation of vibrations. To determine the effect of the cleaning system on the wheel's tendency to load, the flow-rate was measured in relation to coolant pressure for various nozzle cross-sections. Nozzles measuring 0.3–1.0 mm in diameter were used. It was found that nozzles with a bore diameter of <0.6 mm were unsuitable, as they became clogged with chips too quickly. The best results were obtained with nozzle bores of 0.8–1.0 mm at a pressure of 15–20 bar. The flow-rate was 60 l/min.

4.5 CONDITIONING THE GRINDING WHEEL IN HIGH-EFFICIENCY DEEP GRINDING

4.5.1 Dressing and sharpening grinding wheels

When they are delivered, and certainly after they have been in use for a time, grinding wheels are not able to perform their function and/or lose their accuracy of form or profile. This situation is remedied by dressing. Despite many advances in the manufacture of new types of wheel (e.g., with the ability to sharpen themselves when being used), dressing remains an important and cost-intensive process. This is specially true of form wheels for high-performance grinding at high wheel speeds. This process calls for the use of CBN and diamond wheels, which are characterised by their high wear resistance and hardness.

A wheel-dressing process (Fig. 4.6) must fulfil two requirements (**53**)–(**56**).

– It must produce the required accuracy of form and profile, and true-running, of the grinding wheel.
– It must generate a chip space suitable for the grinding process.

Non-rotating dressing tools include single-point and multi-point dressers, and dressing sticks (blocks). Rotating dressers include form

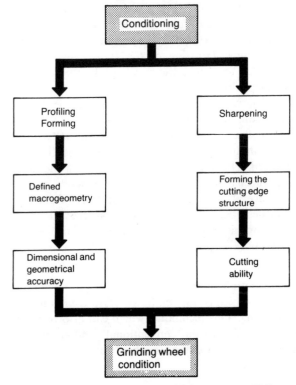

Fig. 4.6 Conditioning and its purpose (54)

wheels, steel rolls, and profile rolls (**54**). The choice of a suitable dressing method and tool is governed firstly by the grinding tool (type of abrasive and bond), and secondly by the technical and economic requirements of the task (**54**).

In the case of diamond and CBN grinding wheels, diamond dressing rolls are used, especially when accurate profiles have to be machined. Wheels with a simple profile, such as cylindrical wheels, are usually dressed with a single-point or multiple-point dresser, or a silicon carbide wheel. Vitrified bond CBN wheels can also be dressed well with non-rotating dressing tools.

At present, no entirely satisfactory methods are known for dressing grinding tools containing ultrahard abrasives. Special difficulties are posed by the small infeed and tolerances of the dressing tool and by the rigidity required of the dressing system, which consists of the dressing tool and the grinding spindle. For a dressing system to be economically viable, it must be capable of reproducible infeeds of the order of 1 μm. A high level of accuracy is especially important for the recently developed TDC process (Touch Dressing of CBN). In this method, infeeds of 2–4 μm are needed to restore the grinding ability of the wheel, without the need for sharpening (**57**)(**58**).

In another, relatively new, technique, the wheel is dressed by milling with a dressing roll, the periphery of which is provided with one or more geometrically defined polycrystalline diamond (PCD) cutting edges (**59**)(**60**). As in milling, the dressing roll engages the grinding wheel without crossfeed. The line of engagement between the dressing roll and the grinding wheel describes a cycloid. At a depth of about 0.05 mm there can be several hundred such curved lines of engagement on the periphery of the wheel. This technique has the advantage that subsequent sharpening is not needed, as the cycloidal lines of engagement achieve a large pore volume, and as a result it is possible to work at a higher specific removal rate.

4.5.2 Swing-step profiling technique

Another method, which is not yet, however, in common industrial use, is swing-step dressing. This technique, which can also be used for dressing form wheels, is different from conventional profiling methods in that it uses an additional axis of swing, i.e., the dressing roll has a total of three possible movements:

–   rotary movement (in both directions if required);

- radial infeed to the grinding wheel;
- swinging of the dressing roll relative to the grinding wheel.

Figure 4.7 shows the principle of the swing-step dressing apparatus, with the facility for measuring dressing forces by means of four 'three-component quartz load cells'. The drive motor powers the rotary movement of the dressing roll. Acting via a gear and a bearing-mounted spindle, a stepping motor enables the dressing roll to be fed in increments to an accuracy of 0.1 μm, while a second stepping motor actuates a swinging movement of around 30 degrees about the centre of rotation.

When a CBN grinding wheel is dressed conventionally with a diamond roll, high forces occur, resulting in undesirable elastic deformation of the dressing system (13). The consequences of these high forces are increased wear on the dressing roll and geometrical errors on the grinding wheel.

In the swing-step process, on the other hand, a modest normal force is generated. As the operation is usually performed with small incremental movements of the dressing tool, the deformation is slight. Reduced deformation results in greater geometrical accuracy and therefore tighter manufacturing tolerances.

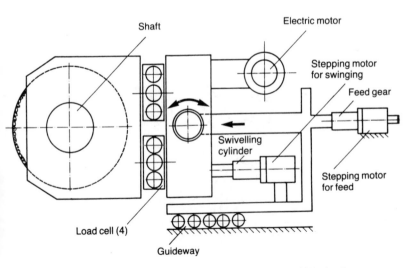

Fig. 4.7   Principle of the swing-step dressing device

At the start of the swing-step dressing process the centre points of the grinding wheel, dressing roll, and axis of swing do not lie on a straight line. The dressing roll is located in its starting position. The dressing roll is fed in this position until it makes contact with the grinding wheel. The rotating dressing roll is then swung at a constant angular velocity. Progressing along the arc of movement, the dressing roll gradually engages the grinding wheel. These relationships are shown in Fig. 4.8. When the centre points of the dressing roll, the grinding wheel, and the axis of swing lie on a straight line, the profiling process is at an end, and the continuing movement of the dressing roll to the other side does not cause further material to be removed from the grinding wheel.

In Fig. 4.8, E_a is the initial point of contact (start of engagement) between the dressing roll and the grinding wheel. As the swinging movement continues, the point of engagement of the dressing roll moves along the arc $\overline{E_a E_e}$. At the end of the dressing process (the final moment of contact between roll and wheel), the dressing roll is in the central position, and point E_e (end of engagement) is in contact. Figure 4.9 shows the relationship between the amount removed by dressing (the dressing

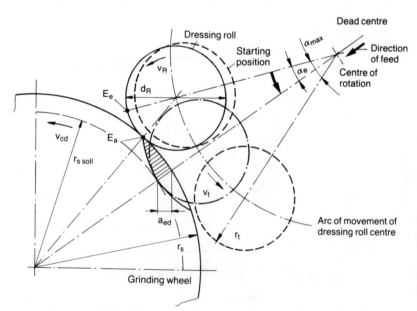

Fig. 4.8 Kinematics of the swing-step dressing process

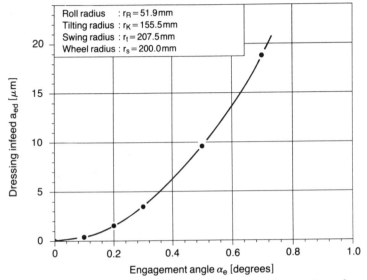

Fig. 4.9   Dressing infeed in relation to engagement angle

infeed) and the corresponding engagement angle α_e for the dimensions indicated. For instance, it is apparent from the graph that, at a dressing infeed of 20 μm, the initial engagement takes place at an engagement angle α_e of 0.7 degrees. As the dressing roll continues to swing, material is removed from the periphery of the grinding wheel. As this is happening, the engagement angle decreases. Figure 4.9 also shows that, if the engagement angle is reduced from 0.7 degrees to 0.5 degrees, the 20 μm dressing infeed is cut by about half. This means that, in swing-step dressing, the specific dressing volume Q'_{wd} increases sharply immediately after the start of engagement, and then slowly falls back to zero. Consequently, a slow decoupling process takes place between grinding wheel and dressing roll. This decoupling may be regarded as an advantage of the swing-step method. The process corresponds to 'spark-out' in grinding, and improves the trueness of the grinding wheel, in addition to promoting the formation of a finer wheel topography.

4.5.3   Touch-dressing CBN wheels (shaving)
Touch-dressing means dressing with minimal infeeds (of the order of microns) in order to splinter the abrasive grains, or the tips of the grains, in such a way that the number of cutting edges engaging the wheel is

increased. This method of conditioning CBN wheels has not yet been fully mastered in industry.

Touch-dressing is mainly used with metal bond, electroplated, and vitrified bond CBN wheels. The result is an abrasive coating with a finer and more uniform surface with which it is possible to achieve high removal rates with relatively uniform surface roughness values (12), (61), (62).

When touch-dressing is carried out on blunted, electroplated CBN wheels, it causes the blunt grains to splinter slightly, thereby restoring the cutting ability of the grinding wheel. Tests have shown that, if the infeeds are too small ($a_{ed} < 2$ μm), the grains do not splinter correctly. But if the infeed is too great ($a_{ed} > 6$ μm), this also adversely affects the grinding process and subsequent sharpening is required (12)(58)(61)(62). This is not needed if the infeed a_{ed} is between 3–5 μm. If these optimum touch-dressing infeeds are used, the results are increased wheel life, reduced variations in grinding forces, and workpieces of more uniform quality.

The main problems in touch-dressing lie in determining the initial contact between the dresser and the wheel, and achieving low infeeds in the 3–5 μm range. Acoustic sensors have now begun to be used to establish this contact (61)(62).

Figure 4.10 is a block diagram of a system which has been developed for detecting initial contact in the touch-dressing of CBN wheels. The noise signal from the dressing tool is picked up by a sensor responding to structure-borne noise, amplified by the charge amplifier and converted to a DC voltage by a diode and a low-pass filter. The DC voltage produced is fed to a Schmitt trigger (comparator) and compared with the latter's reference voltage. When the initial contact occurs, the level of the noise-signal DC voltage rises above the reference voltage of the Schmitt trigger and generates the output signal. This is converted into an audible and light signal, and indicates the initial contact.

It is proposed to connect the output of this system to the control unit of the dressing device or grinding machine in order to determine the further progress of the touch-dressing operation.

The force-measuring sensors used for registering the forces acting on rotating (touch-) dressing tools are able to detect and measure a very small force increase of 1 N. It follows that force-measuring sensors could be used as an alternative to sound sensors for detecting initial contact. Heat and light sensors provide further possibilities for detecting contact.

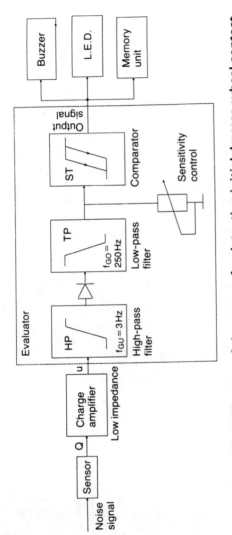

Fig. 4.10 **Block diagram of the system for detecting initial dresser–wheel contact**

Given the rise in temperature in the contact zone, infrared heat sensors, with appropriate signal processing, would be capable of detecting the first contact between the dresser and the wheel.

Another problem posed by touch-dressing is the current lack of information about the optimum tool for this operation. At present, non-rotating, multiple-point dressers are generally used for touch-dressing CBN wheels. These tools wear very quickly, and are not able to maintain accurate infeeds. As they wear at a slower rate, PCD (polycrystalline diamond) inserts are better suited to touch-dressing than multiple-point dressers. Figure 4.11 shows the essentials of a device for touch-dressing (calibrating) the sides of CBN grinding wheels. The tool consists of two round PCD inserts.

More suitable touch-dressing tools would be diamond wheels or rolls, which would wear less and have a much longer tool life.

The essential point is that touch-dressing enables the potential of CBN wheels to be used more fully. Touch-dressing could prolong considerably the tool life of single-layer CBN wheels which have lost their cutting

Fig. 4.11 PCD tools for touch dressing (calibrating) the sides of CBN wheels

ability. In the case of electroplated CBN wheels, a total radial infeed equivalent to about 5–7 percent of the grit size is recommended. The resulting workpiece roughness values are improved by more than 50 percent. However, some further development work is needed here, in particular as regards the detection of contact, the dressing tools, and the rigidity and accuracy of the infeed devices.

4.5.4 Sharpening ultrahard grinding wheels

Sharpening ultrahard grinding wheels is defined as 'freeing', 'opening up' or 'cutting-back' the bond. Sharpening restores the cutting face structure to a form wheel, and, therefore, renews its cutting ability.

The bond can be cut back by various methods. In the usual method, a vitrified bond aluminium oxide block is pressed against the rotating grinding wheel. In practice, these blocks are generally held against the wheel by hand until the bond has been cut back. For reproducible sharpening, apparatus and devices are used which enable a defined feed to be achieved between the sharpening block and the grinding wheel.

Schleich (**63**)(**64**) has investigated this process in depth and has developed the following model for the specific sharpening removal rate Q'_{sb}

$$Q'_{sb} = 0.95 \times q_m^{-1} \times \sqrt{\left\{ C_k \times \left(1 - \frac{R_p}{w_m \times q_m \times z_{crit}} \right) \times R_p^{5/2} \times v_{cd} \right\}}$$

(4.4)

where

q_m = longitudinal elongation factor of the CBN grains ($q_m = 1.41$)
C_k = number of grits per unit volume of the abrasive coating
w_m = average mesh size
z_{crit} = critical grit protrusion
v_{cd} = wheel speed when dressing
R_p = smoothing depth

For practical purposes, the following function can be used for resin and brittle metal bond CBN wheels running at peripheral speeds of $30 \leqslant v_c \leqslant 120$ m/s, and with an average grit protrusion of 30–35 percent of the diameter of the grit (**65**)

$$Q'_{sb} = 0.04 \times \sqrt{(K \times B \times v_{cd})}$$

(4.5)

where

K = concentration (e.g., $K = 0.24$ for V240)
B = grit size (e.g., $B = 151$ for B151 grit)
v_{cd} = wheel speed in m/s while sharpening

Other sharpening techniques include abrasive jet sharpening and chemical erosion. When the former method is used, a blasting medium (loose abrasive) is shot in a suitable carrier (coolants are often used for the purpose) at high pressure against the abrasive surface of the rotating wheel (**66**). Chemical erosion of the bond is used with metal bond diamond and CBN wheels containing medium and coarse grits. In this process the coated portion of the wheel is immersed briefly in an acid. When the desired grit protrusion has been obtained, the grinding wheel is flushed thoroughly. The sharpening operation is carried out away from the machine.

For the sake of completeness, other sharpening processes, such as grinding mild, long-chipping steel, electrical-discharge sharpening, and

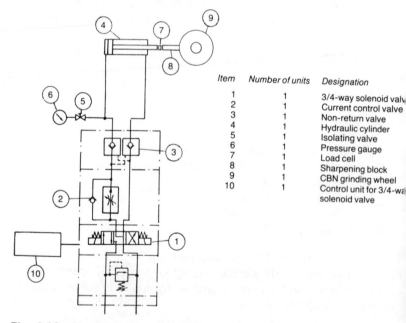

Item	Number of units	Designation
1	1	3/4-way solenoid valve
2	1	Current control valve
3	1	Non-return valve
4	1	Hydraulic cylinder
5	1	Isolating valve
6	1	Pressure gauge
7	1	Load cell
8	1	Sharpening block
9	1	CBN grinding wheel
10	1	Control unit for 3/4-way solenoid valve

Fig. 4.12   The principle of the newly developed sharpening system

the introduction of loose grit into the contact zone, should also be mentioned.

Figure 4.12 shows the principle of a sharpening system for CBN grinding wheels developed recently for the present work. Sharpening blocks are pressed against the grinding wheel at a pressure adjustable up to 1000 N and at a feed rate adjustable up to 100 mm/s. The desired feed rate can be set once for the operation, or in stages, as required, with adjustable increments between the stages. Simultaneous profiling and sharpening results in a reduction in the profiling forces and reduced dressing roll wear, as, during the sharpening process, the bond is cut back and the dressing roll only comes into contact with the abrasive grit.

4.6 TECHNOLOGICAL REQUIREMENTS FOR ACHIEVING HIGH REMOVAL RATES

The cutting process in grinding generates heat and chips which have to be removed from the contact zone between the grinding wheel and the workpiece. As the specific removal rate increases, so too does the process heat and the volume of chips. The grinding process can only be carried out effectively if these are discharged rapidly from the contact zone.

In creep-feed grinding at a relatively low cutting speed and a low grinding output, the amount of energy converted is also fairly small. Because of the relatively large contact length and contact time (low workpiece speed), which are characteristic of this method, some of the energy at low temperature flows into the workpiece. If creep-feed grinding (at a low wheel speed of 30 m/s, for example) is carried out with a high specific removal rate, that is at a high workpiece speed, it is barely possible to dissipate the heat energy adequately, since there is not sufficient time. There is an accumulation of heat, resulting in thermal damage to the workpiece. This is the main reason why conventional creep-feed grinding cannot be carried out with a high specific removal rate.

In high-efficiency deep grinding, even greater quantities of heat are converted in a shorter time. Care must, therefore, be taken to ensure that this energy does not lead to thermal damage to the area of the workpiece close to the surface. With an increasing amount of heat, substantially larger feed rates should be selected in order to prevent the heat from flowing into the workpiece. At high wheel speeds, the grinding wheel, which can also be regarded as a source of heat, leaves the newly created surface more rapidly. The bulk of the energy which is converted into

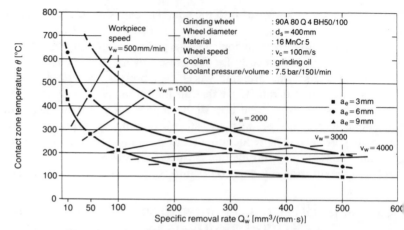

Fig. 4.13 Workpiece surface temperature as a function of specific removal rate

heat is transported away from the contact zone with the chips and the coolant.

Figure 4.13 shows the temperature measured just below the surface of the workpiece as a function of workpiece speed and specific removal rate for various depths of cut (a_e = 3, 6, and 9 mm). This demonstrates that, at a constant depth of cut, the temperature drops as the specific removal rate increases. This is due to an increase in the workpiece speed. As the depth of cut is increased, the surface-zone temperature also increases.

Specific removal rate Q'_w is the product of depth of cut a_e and workpiece speed v_w ($Q'_w = a_e \times v_w$). Raising the depth of cut increases the lengths of contact, which give rise to higher temperatures due to the increasing friction. On the other hand, increasing the workpiece speed leads to a lower grinding temperature. These considerations show that an increase in the specific removal rate is to be achieved primarily by raising the workpiece speed, as the surface temperature of the workpiece then remains lower. A high workpiece speed is, therefore, one of the fundamental prerequisites for high-efficiency deep grinding.

Furthermore, a high wheel speed serves not only to reduce the grinding forces and increase the grinding capacity of the wheel, it also promotes the chip-forming process, and, beyond a certain wheel speed, results in lower workpiece surface temperatures. This relationship is shown in Fig. 4.14.

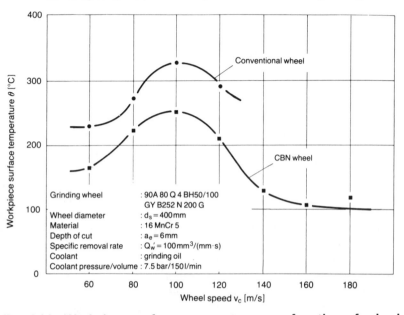

Fig. 4.14 Workpiece surface temperature as a function of wheel speed for CBN and aluminium oxide grinding wheels

This figure shows the temperature curve relative to wheel speed for two different grinding wheels. The wheels in question were an electroplated CBN wheel (B252) and a bakelite bond aluminium oxide wheel suitable for use at high wheel speeds. Up to v_c values of about 100 m/s, the temperature increases with both wheels, although that of the aluminium oxide wheel is at a higher level. When the wheel speed is increased further, the temperature falls back again, attaining at 130 m/s approximately the same level as at 60 m/s. (The aluminium oxide wheel should only be used at wheel speeds up to 120 m/s).

The initial rise in temperature which occurs as the wheel speed increases can be explained by the increase in the frictional energy. As the rotational speed of the wheel increases, less material is removed by each cutting edge engaging the workpiece and the chip thickness is reduced, while the friction becomes greater because the wheel engages the workpiece more frequently. The result is an increase in temperature in the contact zone. This trend continues up to a given wheel speed, and is then reversed.

This drop in surface temperature is explained by the *contact layer theory*. First, the concept of the equilibrium temperature for a cutting edge engaged in the chip-forming process must be considered.

As the cutting edge penetrates the workpiece, both the cross-sectional area of cut and the temperature increase as the length of the cut increases. The cutting-edge temperature only rises to a certain upper limit (the chip-formation equilibrium temperature), thereafter remaining constant over the entire distance of cutting-edge engagement (**15**). In HEDG, this proposition is insufficient to describe the relationships in the topmost layer of the contact area between the grinding wheel and the workpiece, as it refers only to the individual grain; further clarification is required. Figure 4.15(a) shows the curves along which the contact-layer and the contact-zone temperatures tend to move. When the contact-layer temperature reaches the equilibrium point, the maximum surface-zone temperature is also reached. Figure 4.15(b) shows the contact layer greatly enlarged.

To explain the relationships in greater detail, we have to look more closely at the contact layer and its characteristic features. The topmost layer below the area of contact between the wheel and the workpiece may be defined as the 'contact layer'. Its thickness corresponds to the depth of cut (Fig. 4.15).

It is generally the case that in HEDG the kinematic number of cutting edges N_{kin} is greater than in creep-feed grinding. The number of instantaneously engaged cutting edges is proportional to the kinematic number of cutting edges. In HEDG this results in a greater number of cutting paths which lie very close together, as shown in Fig. 4.16. As described above, the engagement of each cutting edge causes a high temperature after a short build-up time. Owing to the high number of engagements and the effect of thermal conductivity, the entire contact layer is heated to temperatures of 1000–1800°C and reaches the equilibrium temperature.

Stähli (**67**) has observed that heat initially spreads more quickly over the surface than into the core of steel components. He heated a steel surface to melting point using high-frequency electron beam pulses. The contact area measured 1 mm in diameter and was heated for 11.1 ms (pulse duration) at 200 W/mm². Because of the high thermal conductivity of the workpiece, at first the heat flows more rapidly over the surface than into the core, due to the lower thermal conductivity of the adjacent layer of air. Figure 4.17 shows the isotherms for the spread of heat over the surface and into the core of the workpiece.

(a)

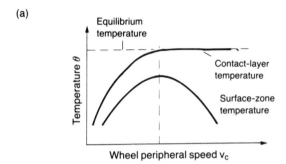

(b)

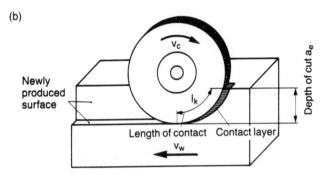

Fig. 4.15 **(a)** **Temperature in the contact layer and surface zone relative to grinding wheel peripheral speed**
(b) **Diagram showing the contact layer in the contact zone**

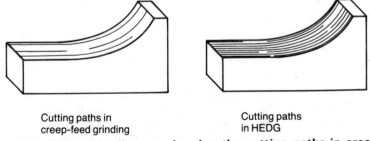

Cutting paths in
creep-feed grinding

Cutting paths
in HEDG

Fig. 4.16 **Schematic diagram showing the cutting paths in creep-feed grinding and HEDG**

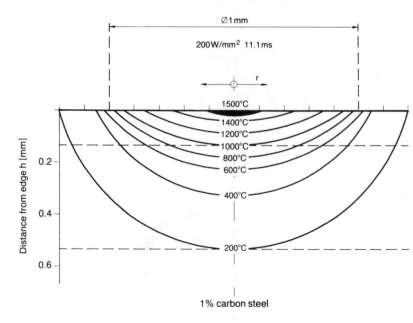

Fig. 4.17 **Temperature distribution resulting from brief thermal loading, measured immediately after heating has ceased (after (67))**

The effect observed by Stähli is relevant to high-efficiency deep grinding. If the introduction of heat resulting from the engagement of a cutting edge is equated with the brief electron pulse, it can be inferred that the heat build-up due to the engagements of the cutting edge is greater at the sides than in the core of the workpiece. The lateral heat distribution facilitates the cutting process for the adjacent cutting edge, so reducing the grinding forces. In creep-feed grinding this situation does not arise, as the cutting paths are more widely separated owing to the lower kinematic number of cutting edges.

The time required for the contact layer to be removed can be determined mathematically, as shown below by way of example. Given a depth of cut a_e of 6 mm, a specific removal rate Q'_w of 100 mm^3/(mm · s) and an average wheel speed v_c of 100 m/s, the time required to remove the contact-layer can be calculated as follows.

As a close approximation, the length of contact l_k is

$$l_k = \sqrt{(a_e \times d_s)} \tag{4.6}$$

For a depth of cut a_e of 6 mm and a grinding wheel diameter d_s of 400 mm, l_k is given by the following expression

$$l_k = \sqrt{(6 \text{ mm} \times 400 \text{ mm})} \approx 49 \text{ mm}$$

The equivalent thickness of the contact layer h_{eq} can be determined by the following equation

$$h_{eq} = \frac{a_e \times v_w}{v_c} = \frac{Q'_w}{v_c} \tag{4.7}$$

$$h_{eq} = \frac{100 \text{ mm}^3/(\text{mm} \cdot \text{s})}{100 \text{ m/s}} = 0.001 \text{ mm}$$

The volume of the contact layer, with a width b_k of 1 mm, is calculated as follows

$$V'_{sch} = l_k \times h_{eq} \tag{4.8}$$

$$V'_{sch} = 49 \text{ mm} \times 0.001 \text{ mm} = 0.049 \text{ mm}^3/\text{mm} \tag{4.9}$$

The relationship between the specific removal volume for the contact layer V'_{sch} and the specific removal rate Q'_w gives the time t_{sch} in which the contact layer is removed

$$t_{sch} = \frac{V'_{sch}}{Q'_w} = \frac{0.049 \text{ mm}^3/\text{mm}}{100 \text{ mm}^3/(\text{mm} \cdot \text{s})} = 0.49 \text{ ms} \tag{4.10}$$

The contact layer is formed anew continuously, about 2000 times a second. Because of these relatively short times, the heat barely has a chance to flow into the workpiece. The grinding energy is removed from the contact zone primarily with the contact layer in the form of chips.

The heat flowing into the workpiece via the contact layer depends on two factors: the thickness of the layer and the temperature. The product of the volume of the layer (varying depending on its thickness) and the specific heat of the workpiece material is the heat content of the contact layer.

At the reversal point (Fig. 4.14), high temperatures in the contact layer facilitate the chip-forming process. When this point is passed, stepping up the wheel speed produces no further increase in frictional energy, but accelerates the removal of heat from the contact zone. In high-efficiency deep grinding, a zone is, therefore, formed in the upper contact layer where the equilibrium temperature (about 1000–1800°C) prevails. From this point on, the thickness of the contact layer diminishes as the peripheral speed of the grinding wheel increases, while the (equilibrium) temperature remains unchanged. The thinner this layer becomes, the less energy it is able to absorb despite the high temperature level. There is a fall in temperature in the zone below the thin contact layer and on the newly produced surface. At lower wheel speeds the contact-layer temperature is below the equilibrium temperature (see Fig. 4.15), and the surface-zone temperature, therefore, remains low in spite of the greater thickness of the contact layer. The reason for this is that the contact layer is removed before the temperature of the contact zone is propagated to the underside of the contact layer.

An increase in specific removal rate does not result in a proportionate increase in grinding forces and grinding output.

In HEDG the specific energy requirement is much lower than in conventional creep-feed and reciprocating grinding. In some cases it is less than 10 percent of that required for conventional grinding. The main reasons for this are as follows.

(a) The proportion of the cutting energy needed for the elastic deformation of the workpiece is lower in HEDG than in conventional grinding because the hot contact layer deforms more readily and more rapidly.

(b) Because of the higher speed at which deformation takes place and the high temperature of the contact layer in HEDG, lower forces and energy are required for the plastic deformation of the workpiece in the chip-forming process.

(c) A further reason for the lower specific energy requirement is the greater chip thickness due to larger depths of cut and increased feed rates. In general, less energy is required if a given chip volume is removed in one cut, than if it is removed in several thinner layers. Compared with creep-feed grinding, the chip thickness in HEDG is considerably greater owing to the higher workpiece speed.

If a machining operation is performed with unsuitable tools, grinding parameters, dressing conditions, and/or an inadequate supply of coolant, the process is likely to be seriously impaired. If the grinding wheels are loaded, friction increases sharply, leading to an increased build-up of heat in the contact zone, and so to a high thermal load on the workpiece, possibly resulting in damage.

In the case of materials which are difficult to machine, such as titanium, the frictional element in the chip-forming process is very high. Consequently, the process demands greater energy, which cannot be removed from the contact zone without thermal damage. The difficulty can usually be remedied by a lower energy input, for example lower cutting power and a lower wheel speed **(68)(69)**.

Figure 4.18 shows the qualitative effect of workpiece speed and specific stock removal rate on the surface-zone temperature. If the process is set up correctly, the surface-zone temperature reached in creep-feed grinding lies in ranges where the workpieces do not suffer any damage **(4)**. Between the creep-feed grinding range and the HEDG range there is a zone characterised by higher temperatures. The high temperature in this critical range, which lies just beyond the conventional creep-feed range, is one of the factors responsible for the failure of attempts to achieve higher specific removal rates using the creep-feed technique.

A high removal rate simultaneously implies a large chip volume, which has to be accommodated in the chip-spaces of the grinding wheel and removed from the contact zone with the coolant. The chips produced in

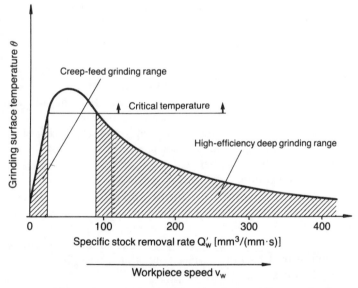

Fig. 4.18   Effect of workpiece speed and specific removal rate on surface-zone temperature

one revolution of the grinding wheel must have a volume less than that of the chip-spaces on the periphery of the wheel. It follows that the chip spaces dictate the limit for the maximum specific stock removal rate. The chip-spaces are increased by using coarser grit or by making the pore spaces larger, although this may have adverse effects in other respects.

Apart from producing a rougher surface, coarser grit gives rise to higher grinding forces, and also the number of cutting edges engaging the workpiece is reduced. In addition, increasing the pore spaces reduces the mechanical strength of the abrasive layer, and this is not always desirable.

A higher grit concentration is another factor which serves to promote higher specific removal rates, as it results in a higher number of kinematic cutting edges and lower chip thickness. In general, higher grinding forces must be expected. In high-efficiency deep grinding, the grinding forces do not increase proportionally with the specific stock removal rate, and remain relatively low. A higher concentration combined with a large chip-space volume can be achieved with electroplated wheels and this is why such wheels are often the most suitable tools for HEDG.

The coolant also plays an important part in the chip-forming process, lubricating the contact zone and removing heat and chips. Good lubricating and cooling effects are required at the same time, although in practice this is not possible. Grinding oils have a good lubricating effect while water-soluble lubricants are better at absorbing heat. For certain materials and machining conditions the dissipation of heat is more important, while in many other operations the lubricating action is the more important function. In high-efficiency deep grinding, oil, with its superior lubricating properties, has established itself as the most widely used coolant (12)(13). Optimum coolant supply and wheel cleaning are of great importance in HEDG, and are considered in section 4.4.

In summary, it can be said that the contact layer, as defined above, provides an explanation for the factors at work in HEDG which lead to reduced surface-zone temperatures in the workpiece. The rapid removal of the hot contact layer is effected by the high speed of the wheel and the workpiece, which is a fundamental requirement for high-efficiency deep grinding. The rapid removal of the contact layer prevents any major transmission of heat into the surface zone, thus preventing any thermal damage to the workpiece.

Apart from the need for a suitable machine, a high wheel speed, a large depth of cut, and a high workpiece speed, these theoretical con-

iderations and empirical findings indicate that high removal rates can
>e achieved with:

- electroplated CBN grinding wheels (because of their capacity for
  operating at high wheel speeds);
- medium-size grit;
- microcrystalline grit, because of its favourable fracturing properties
  (The splitting away of small fragments restores the sharpness of the
  grains with little wear (**61**));
- an oil coolant, with suitable supply systems and pressures.

In high-efficiency deep grinding, the specific removal rate should be
increased by raising the workpiece speed, rather than by increasing the
depth of cut (see Fig. 4.13).

4.7 ANALYTICAL DETERMINATION OF THE WHEEL-SPECIFIC MAXIMUM REMOVAL RATE

As with conventional reciprocating and creep-feed grinding, there are
various factors which impose performance limits on high-efficiency deep
grinding. Figure 4.19 summarises the most important limiting factors
which affect HEDG. These are presented in the form of the quantities

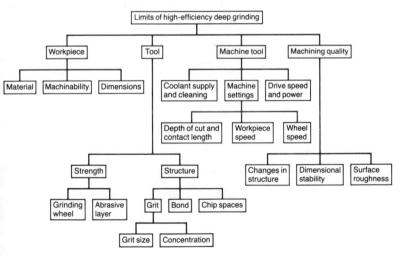

Fig. 4.19 Limiting factors in high efficiency deep grinding

and characteristics associated with the workpiece, the grinding wheel, the grinding machine, and the machining quality. The requirements which apply in assessing these factors, as they relate to the grinding machine, the grinding wheel, machine settings, etc., have been described in sections 4.1, 4.2, and 4.3.

In this section two key limiting factors are considered in more detail: chip space, and the mechanical strength of the wheel. These considerations are based on an analytically derived model for determining the limit value of the specific stock removal rate.

4.7.1 Mean chip space as a limiting factor

The function of the chip spaces on the grinding wheel is to receive the material removed from the workpiece and transport it away from the contact zone. Essentially, chip spaces are created by cutting back the bond in the sharpening operation. The chip space volume on CBN wheels can be determined mathematically or by taking traces of the wheel surface. An analytically derived mathematical model for determining the maximum possible specific stock removal rate is described below.

To calculate the chip space, a geometrically definable ideal grain must be postulated, arranged in a particular manner in the bond and on the substrate. CBN grains of octahedral shape are taken as the model, as the scanning electron microscope shows that this is the most commonly occurring shape (**23**)(**63**)(**70**). Figure 4.20 shows idealized grains schematically arranged on the substrate.

The figure illustrates 'critical' grit protrusion (i.e., where the grit is not solidly anchored in the bond). This value is $d_k/2$, at which value the bond has been cut back to half the size of the grit or, conversely, covers up to

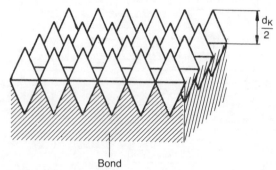

Bond

Fig. 4.20   Idealized grit, and its arrangement on the substrate

50 percent of the diameter of the grit. The volume of the individual grain V_k may be determined with equation (4.11)

$$V_k = \tfrac{1}{6} \times d_k \times w_m^2 \tag{4.11}$$

where w_m is the average mesh size of the grit.

Half the individual grain volume above the level of the bond is given by the expression

$$\frac{V_k}{2} = \frac{1}{6} \times \frac{d_k}{2} \times w_m^2 \tag{4.12}$$

The free space above the centreline of the grain (the bond level) corresponds to the chip space V_{sp} of an individual grain

$$V_{sp} = V_{ges} - \frac{V_k}{2} \tag{4.13}$$

where V_{ges} represents the total volume of a half-grain including the corresponding chip space

$$V_{ges} = \frac{1}{2} \times \frac{d_k}{2} \times w_m^2 \tag{4.14}$$

The individual chip space V_{sp} can be determined by inserting equations (4.12) and (4.14) in expression (4.13)

$$V_{sp} = \left(\frac{1}{2} \times \frac{d_k}{2} \times w_m^2 \right) - \left(\frac{1}{6} \times \frac{d_k}{2} \times w_m^2 \right) \tag{4.15}$$

$$V_{sp} = \tfrac{1}{6} \times d_k \times w_m^2 \tag{4.16}$$

If all the grains are arranged in idealized form next to each other on the periphery of a cylindrical wheel of diameter d_s and width b_s, the number of grains n_k can be calculated as follows

$$n_k = \frac{2\pi \times d_s \times b_s}{w_m^2} \tag{4.17}$$

The sum of all the chip spaces on this grinding wheel with 50 percent grit protrusion is then given by the expression

$$V_{sp,\,ges} = \tfrac{1}{6} \times d_k \times w_m^2 \times n_k = V_{sp} \times n_k \tag{4.18}$$

From equations (4.17) and (4.18) it follows that

$$V_{sp, ges} = \tfrac{1}{6} \times d_k \times w_m^2 \times \frac{2\pi \times d_s \times b_s}{w_m^2} \qquad (4.19)$$

In equation (4.19) the mesh size w_m and the average grit diameter d_k are given in mm.

After transformation, the following is obtained

$$V_{sp, ges} = \tfrac{1}{3} \times d_k \times \pi \times d_s \times b_s \qquad (4.20)$$

Table 4.2 shows, for five different grit sizes, the number of grains and chip-spaces per millimetre of wheel width on a grinding wheel with a diameter d_s of 400 mm. The average grit diameter and the average mesh size are as indicated in the FEPA standard and DIN 848.

Taking all the chip spaces into account, the limit for the maximum specific removal rate $Q'_{w, max}$ can be calculated by the following equation

$$Q'_{w, max} \leq V_{spl} \times n_s \qquad (4.21)$$

where n_s is the rotational speed of the grinding wheel.

The maximum specific removal rate for grit sizes B252 and B151, and a wheel speed v_c of 180 m/s can be calculated, assuming

grinding wheel diameter $d_s = 400$ mm
rotational wheel speed, $n_s = 8592$ min^{-1} ($= 143.2$ s^{-1}),
which corresponds to a wheel speed, v_c of 180 m/s

$$Q'_{w, 252} = 105.556 \text{ mm}^3/\text{mm} \times 143.2 s^{-1} = 15,115.6 \text{ mm}^3/(\text{mm} \cdot \text{s})$$
$$(4.22)$$

for B252 grit

$$Q'_{w, 151} = 63.249 \text{ mm}^3/\text{mm} \times 143.2 \text{ s}^{-1} = 9057.2 \text{ mm}^3/(\text{mm} \cdot \text{s}) \qquad (4.23)$$

for B151 grit

These values are theoretical limits, as they cannot be achieved in practice owing to factors which prevent the chip space from being fully utilized. For instance, the chips require more space than their actual volume. If the chip space were equal to the chip volume, a large quantity of energy would be required to work the chips.

For conventional grinding methods and wheels, a degree of chip space filling, ξ of 0.2 is regarded as the optimum. In high-efficiency deep grind-

Table 4.2 **Various characteristics and mesh sizes, and the maximum possible chip space per grain, for a cylindrical wheel measuring 400 mm in diameter (d_s) and 1 mm in width (b_s)**

Average grit diameter d_k (μm)	Average mesh size w_m (μm)	Number of grains per mm of wheel width n_k	Chip space per individual grain v_{sp} (mm^3)	Chip space per mm of wheel width v_{spl} (mm^3)
252	215	54 370	1.9412×10^{-3}	105.556
181	165	92 314	0.8212×10^{-3}	75.812
151	137	133 905	0.4723×10^{-3}	63.249
126	117	183 598	0.2874×10^{-3}	52.775
64	58	747 108	0.0358×10^{-3}	26.783

ing, the short deformation times allow somewhat higher chip space filling coefficients ($\xi = 0.2$–0.3).

The degree of chip space filling is the quotient of the individual chip volume V_{cu} and the corresponding chip space volume V_{sp} **(23)(49)(63)(70)**

$$\xi = \frac{V_{cu}}{V_{sp}} \tag{4.24}$$

Another factor is the influence of grit protrusion. In the proposed model, and in Table 4.2, a grit protrusion z of $1/2\, d_k$ is assumed. Because of the danger of breakout, this protrusion is critical, and should be avoided wherever possible. A grit protrusion of $z = 0.35$–$0.45\, d_k$ has proved most suitable for HEDG. The chip spaces with a smaller grit protrusion ($z < d_k/2$) can be calculated from the following equations, used in conjunction with Fig. 4.21.

$$\frac{b_z}{w_m} = \frac{z}{d_k/2} \tag{4.25}$$

from which it follows that

$$b_z = \frac{2w_m \cdot z}{d_k} \tag{4.26}$$

The chip space volume V_{spb} which is no longer available because of the reduced grit protrusion $z < d_k/2$ can be calculated for individual grains

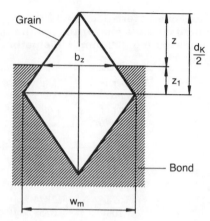

Fig. 4.21 Different grit protrusions and associated geometrical quantities

by applying the following equation

$$V_{spb} = V_{sp} - (V_{spz} - V_{bz}) \tag{4.27}$$

where

$$V_{bz} = \tfrac{1}{6} \times z \times b_z^2 \tag{4.28}$$

$$V_{spz} = \tfrac{1}{2} \times z \times w_m^2 \tag{4.29}$$

With V_{sp} obtained from equation (4.13) and b_z from equation (4.26)

$$V_{spb} = (\tfrac{1}{6} \times d_k \times w_m^2) - (\tfrac{1}{2} \times z \times w_m^2) + \left(\tfrac{2}{3} \times z \times \frac{w_m^2 \times z^2}{d_k^2} \right) \tag{4.30}$$

This equation can be simplified to

$$V_{spb} = \tfrac{1}{6} \times d_k \times w_m^2 \times \left(1 - \frac{3z}{d_k} + 4 \times \frac{z^3}{d_k^3} \right) \tag{4.31}$$

For different grit protrusions z, equation (4.31) gives the following values for V_{spb}

$$z = 0.5 \quad V_{spb} = 0 \tag{4.32}$$

$$z = 0.45 \quad V_{spb} = 0.014 \times V_{sp} = 1.4\% \times V_{sp} \tag{4.33}$$

$$z = 0.35 \quad V_{spb} = 0.121 \times V_{sp} = 12.1\% \times V_{sp} \tag{4.34}$$

Where the grit protrusion is different, allowance has to be made for a grit protrusion factor K_z

$$K_z = \frac{V_{sp} - V_{spb}}{V_{sp}} \tag{4.35}$$

If a grit protrusion z of 0.35–0.45 d_k is assumed, the following value for K_z applies

$$K_z = 0.98\text{–}0.88 \tag{4.36}$$

The chip space is reduced further if the bond displacement factor is taken into consideration. The bond is not removed or eroded uniformly in front of and behind the grain. Behind the grain some of the bond remains in the form of a ridge, and this reduces the available chip space. Figure 4.22 shows the ridges of bond behind the grains. These ridges can occupy up to 30 percent of the chip space. This means that values of 0.7–1.0 have to be assigned to the bond displacement factor B_b.

The other factors which inhibit the full utilization of the chip space, e.g., the material, the shape of the chip, the type of bond, the different coefficients of friction between the chip and the bond in the chip space, the grit size and its concentration on the grinding wheel, the vibration of the grinding machine, and the coolant, are subsumed in a general factor f_0. This factor can range from 0.1 to 1. Under optimum conditions, and with materials possessing good grinding properties, $f_0 = 1$.

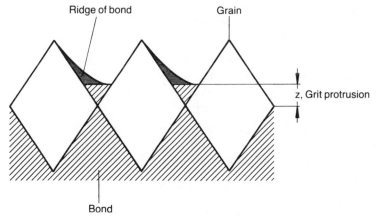

Fig. 4.22   Ridges of bond behind the grains

Taken together, all the determining factors can be translated into the chip space reduction factor f_1. The latter can be calculated as below, where f_0 is the general factor, ξ the chip space filling coefficient, B_b the bond displacement factor, and K_z the grit protrusion factor

$$f_1 = f_0 \times \xi \times [1 - \{(1 - B_b) + (1 - K_z)\}] \tag{4.37}$$

where

$$f_0 = 0.1\text{--}1.0$$
$$\xi = 0.20\text{--}0.3$$
$$B_b = 0.70\text{--}1.0$$
$$K_z = 0.88\text{--}0.98$$

The minimum and maximum values for the reduction factor f_1 can be determined by substituting the minimum and maximum values for f_0, ξ, B_b, and K_z, producing the following rounded values for f_1

$$f_1 \stackrel{.}{=} 0.012\text{--}0.294 \tag{4.38}$$

By allowing for the reduction factor f_1, it is possible to calculate the total chip volume, or the maximum volume of stock, which can be removed in one revolution of the grinding wheel

$$V_{spv} = f_1 \times V_{sp, ges} \tag{4.39}$$

from which it follows that

$$V_{spv} = f_1 \times \tfrac{1}{3} \times d_k \times \pi \times d_s \times b_s \tag{4.40}$$

If the chip volume which can be accommodated on the periphery of the grinding wheel is multiplied by the number of revolutions per second, the result is the limit value for removal rate, which is equal to the specific removal rate per mm of wheel width.

$$V'_{spv} = V_{spv} \times n_s \tag{4.41}$$

The following equation is also valid

$$V'_{spv} = Q'_{w, max} = \tfrac{1}{3} \times f_1 \times d_k \times \pi \times d_s \times n_s \tag{4.42}$$

The expression $(\pi d_s n_s)$ corresponds to the wheel speed v_c, and is used in equation (4.42). This gives the cutting performance which can be achieved with electroplated CBN wheels

$$V'_{sp} = Q'_{w, max} = \frac{1}{3} \times f_1 \times d_k \times v_c \qquad (4.43)$$

Specimen calculation

For two 400 mm diameter CBN grinding wheels containing B151 and B252 grit, the following limit values for stock removal rate are obtained, taking into consideration the reduction factor f_1

$$Q'_w = 181.4 - 4445 \text{ mm}^3/(\text{mm} \cdot \text{s}) \text{ for B252 grit}$$

$$Q'_w = 108.7 - 2255 \text{ mm}^3/(\text{mm} \cdot \text{s}) \text{ for B151 grit}$$

With electroplated CBN grinding wheels used on material with relatively good grinding properties under optimised grinding conditions, the following specific stock removal rates were obtained in the test

for B252: $\quad Q'_w = 3000 \text{ mm}^3/(\text{mm} \cdot \text{s})$

for B151: $\quad Q'_w = 2000 \text{ mm}^3/(\text{mm} \cdot \text{s})$

These results are just below the upper limit of the mathematically established values. The analytical relationship presented here is restricted to high-efficiency deep grinding, primarily with electroplated CBN wheels.

4.7.2 The strength of the wheel body and bond as a limiting factor

Steel, aluminium, aluminium–resin, and carbon-fibre reinforced plastic are the only suitable materials for the wheel body for high-efficiency deep grinding at wheel speeds $v_c > 120$ m/s. Substantial stresses due to clamping forces, grinding forces, and centrifugal forces occur in the wheel body during grinding. The centrifugal forces are critical here, with the maximum tangential stress σ_{tmax} at the edge of the bore (38)(39).

In determining these stresses, the grinding wheel may be regarded, in simplified form, as a rotating disc with a borehole passing through its centre. For cylindrical wheels of uniform thickness, the maximum tangential stress can be calculated by applying the following equation (39)(71)

$$\sigma_{tmax} = \rho \times \omega^2 \times \frac{3 + v}{4} \times r_a^2 \times \left(1 + \frac{1 - v}{3 + v} \times \frac{r_i^2}{r_a^2}\right) \qquad (4.44)$$

For a steel wheel-core (with Poisson's ratio $v = 0.3$, density ρ, angular velocity ω, outside radius r_a, and inside radius r_i), equation (4.44) can be simplified to

$$\sigma_{tmax} = 0.825 \times \rho \times \omega^2 \times r_a^2 \times \left(1 + 0.212 \times \frac{r_i^2}{r_a^2}\right) \qquad (4.45)$$

Because of the different physical and chemical characteristics of the various abrasive coatings, adhesion problems can arise at the joint with the wheel body. Consideration of all the phenomena occurring at this joint does not form part of the present study. The adhesion problem could be reduced if chemical bonding of the grit, the bond, and the wheel body were possible.

To determine the tangential stresses in the abrasive coating, this may be regarded as a thin-walled rotating disc. The stresses can then be calculated by applying equation (4.46)

$$\sigma_t = \sigma + \omega^2 \times r_a^2 \qquad (4.46)$$

The maximum wheel speeds for some wheel bodies and grinding wheels have been determined in tests:

- resin bond grinding wheels or wheel bodies: $v_c \leqslant 120$ m/s;
- vitrified bond grinding wheels or wheel bodies: $v_c \leqslant 120$ m/s;
- aluminium/resin, aluminium and steel grinding wheels and wheel bodies: $v_c \leqslant 250$ m/s in each case;
- special wheel design and material: $\leqslant 500$ m/s.

The bursting strength of these latter materials is generally well in excess of the functional strength required. However, it is good practice not to load the wheel body to a point close to its bursting strength, as it suffers unacceptable deformation before the bursting speed is reached.

The strength of the adhesive joints on segmented grinding wheels limits the use of adhesives as cementing or bonding agents. At present grinding wheels of this type can be used at peripheral speeds v_c of up to 120 m/s. The further development of suitable adhesives or improved fastening methods may lead to the increased use of segmented wheels in high-efficiency deep grinding.

The strength of the bond is critical to grit retention, i.e., as the strength of the bond increases, the loading on the grit can also increase. Yegenoglu (23) has investigated this relationship for resin bond CBN wheels. In a test he used a sharp implement to dislodge the grains from the bond, and measured the forces involved. This grain breakout force increases with increasing grit size, and decreases with increasing smoothing depth R_{ps} (Fig. 4.23). An increase in smoothing depth results in greater grit protrusion. At a smoothing depth R_{ps} of 40 μm, Fig. 4.23 shows the following grain breakout force for two different grit sizes

$F_{kmax} = 32$ N for B252

$F_{kmax} = 16$ N for B151

As resin bond CBN wheels are not the optimum tools for high-efficiency deep grinding, for the most part the tests were performed with metal bond (single and multi-layer) wheels. Because of their different mechanical strengths, it is of little value to make a direct comparison between resin and metal bonds. The grit-retention forces for resin bond wheels (Fig. 4.21) may, however, be used as reference values. With metal bond wheels, the greater mechanical strength of the bond suggests that the grain breakout forces will in any case be greater. In other words, if the grain breakout force for a resin bond is applied to a metal bond, the discrepancy will be on the side of safety.

Table 4.3 shows the grinding forces arising in high-efficiency deep grinding using two electroplated CBN wheels with different grit sizes.

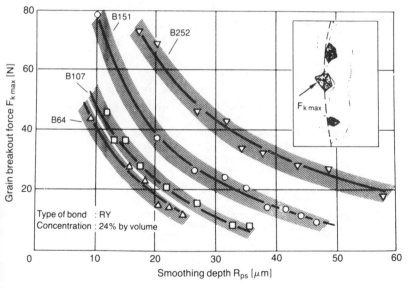

Fig. 4.23 The loading capacity of resin bond CBN wheels relative to grit size and smoothing depth (23)

Table 4.3 Grinding forces per unit area for two electroplated CBN grinding wheels, and grinding parameters

CBN grit	Grinding forces per unit area			Operating parameters
	F_n'' (N/mm^2)	F_t'' (N/mm^2)	F_{ges}'' (N/mm^2)	Material: 16MnCr5 Wheel speed, $v_c = 180$ m/s
252	9.09	5.54	10.64	Spec. removal rates:
151	8.54	4.77	9.78	$Q_{w252}' = 3000$ mm^2/(mm · s) $Q_{w151}' = 2000$ mm^2/(mm · s)

The measurements of the grinding force components per unit area can be used to calculate the mechanical strength of the bond, and therefore the specification of the grinding wheel. The calculation of the grinding forces per unit area was based on the extremely high specific removal rates of $Q_w' = 3000$ mm^3/(mm · s) for B252 grit, and 2000 mm^3/(mm · s) for B151 grit.

To determine the relationship between the grain breakout force and the forces acting on the individual grains in high-efficiency deep grinding, it is necessary to establish the number of grains per mm^2 of wheel area.

Table 4.4 shows the calculated number of CBN grains per mm^2 of wheel for an ideal arrangement of grains, together with the actual number of grains on the abrasive layer of the grinding wheels used in the test. A grain count of wheels from two different toolmakers revealed only slight differences for electroplated CBN wheels.

Dividing the specific grinding forces for a given grit size from Table 4.3 by the counted number of grains/mm^2 (Table 4.4) gives the force acting

Table 4.4 Calculated and counted numbers of grains per mm^2 on electroplated CBN wheels of different grit sizes

Grain diameter for CBN grit d_k (μm)	Number of grains per mm^2, calculated for ideal grain arrangement	Grains counted on abrasive layer
427	13	6.5–7
252	43	16–17
181	68	–
151	99	23–25
126	136	–

on a single grain. This assumes that all the grains engage the workpiece. However, even if only one grain in twenty engages, the forces acting on a single grain in electroplated wheels (at an extremely high Q'_w value) are still lower than the grain breakout force for resin bond grinding wheels. From this it may be concluded that, in high-efficiency deep grinding, the mechanical strength of the bond is of secondary importance as a limiting factor for maximum specific stock removal rate.

On the other hand, the mechanical strength of the bond is vitally important to the service life of a grinding wheel, in terms of the erosion of the bond by the chips. The erosion of the bond is determined, among other things, by the properties of the workpiece material, the shape of the chip, and the degree of filling of the chip spaces. Clarification of the exact relationships between these factors would demand extensive research.

The limiting factors applicable to high-efficiency deep grinding may be summarised as follows: for CBN grinding wheels with suitable (aluminium or steel) wheel bodies, the chip space capacity is the principal factor determining the maximum possible specific stock removal rate. The other factors, such as the mechanical strength of the bond, are of secondary importance.

4.8 THERMOMECHANICAL PROCESS CONDITIONS

The limiting factor in grinding operations is usually the quantity of heat flowing into the workpiece, as this can lead not only to burning, micro-cracking, skin-softening, and geometrical and dimensional errors, but also to undesirable residual tensile stresses in the surface of the work-piece.

The heat generated in grinding is due to external and internal friction arising from elastic and plastic deformation, as well as to shearing and cutting actions (**9**). The more rapidly this heat is removed from the contact zone, the more favourable is the effect on high-efficiency grinding. Grinding temperature is governed by various factors and parameters, which are listed in Table 4.5 under their appropriate headings.

The effects of these factors vary. Some have a profound, some a lesser, influence on heat generation. It must also be borne in mind that the effect of some factors may vary in response to other factors. It would be an extremely exacting task to determine accurately the effects of all these factors. In the grinding experiments described here, the variations are, therefore, confined to those parameters relevant to normal practice, while the others are kept constant.

Table 4.5 Factors and parameters affecting grinding temperature

Coolant and wheel-cleaning	*Grinding wheel*
Supply	Type of grit
Composition	Grit size
Concentration	Bond
Flow-rate	Pore spaces
Pressure	Concentration
Cleaning	Topography
Quantity of cleaning agent	Dimensions
Pressure of cleaning agent	State of balance
Additives	Tendency to load
	Wear
	Roundness
	Modulus of elasticity
Workpiece	*Grinding and dressing parameters*
Chemical composition	Depth of cut
Structure	Workpiece speed (feed)
Hardness	Wheel speed
Geometry of the workpiece	Grinding method
Clamping of the workpiece	Grinding direction
Chip shape	Dressing conditions
Machining properties	
Thermo-physical properties	

In addition, the measuring of temperature in the contact zone presents great difficulties. Many researchers have tried to measure grinding temperature using thermocouples **(9)–(11)(72)–(74)**. Single and two-wire thermocouples, both sheathed and unsheathed, have been used for this purpose. However, such efforts have generally led to widely divergent results.

The main problem in measuring the temperature in the contact zone is due to the rapid and irregular chip forming process. Grinding involves deformation rates of 10^5–10^9 s^{-1} (10^4–10^6 s^{-1} in turning and 10^2–10^5 s^{-1} in hobbing) **(75)**. The high speed of deformation, the brevity of the measuring period, and other factors such as the coolant, limit the opportunity for accurately determining the temperature in the contact zone, and during the chip forming process. In spite of all the difficulties, however, this temperature must be measured for research purposes and has to be ascertained by new and less awkward methods. Most authors who have worked on this have carried out their measurements in conventional reciprocating grinding (with a high workpiece speed and small depth of cut) (9)–(11)(72)(74)(76)(78).

To date, very few measurements have been carried out in creep-feed grinding, a method in which the thermocouples have sufficient time to respond to the heat flow because of the relatively slow speed of the workpiece. In high-efficiency deep grinding, where a greater amount of energy is converted in a shorter time compared with other grinding methods, it is very important to know the temperatures prevailing in the contact zone and on the newly produced surfaces.

The question becomes of even greater interest when considering previous studies which come to the conclusion that the quantity of heat increases steadily with rising wheel speed and increasing specific stock removal rate. In HEDG, these values are many times higher than those encountered in conventional grinding. Furthermore, no satisfactory explanation has yet been given for the behaviour of ultrahard CBN wheels compared with aluminium oxide wheels, and the temperatures to which they give rise in the workpiece.

4.8.1 Theoretical principles and calculation methods

A number of cutting edges of the abrasive grains engage the material in the contact zone between the grinding wheel and the workpiece and cause material to be removed. Each cutting edge involved in the process removes material from a small area (of a few microns) in a very short time and at high speed. Chip formation is irregular and is associated with the conversion of mechanical energy into heat.

The chip-forming process can be broken down into the following stages. Initially, the grain causes the material to undergo elastic deformation. As the cutting edge continues to move through the workpiece, the mechanism of elastic deformation becomes one of plastic deformation with internal friction.

Finally, the effects of external elastic and plastic deformation and of internal friction within the workpiece result in the formation of the chip (8). Mechanical energy is converted into heat and is discharged from the contact zone into the chips, the coolant, the grinding wheel and the outer layer of the workpiece close to the surface.

The total converted heat flow Q cannot be equated with the power output of the spindle P_c. Part of the output is lost by the generation of an air flow, the transport of the coolant, the transport of the chips and the mechanical friction of the spindle, and is not converted into heat in the contact zone. These energy components are taken into consideration by the heat conversion factor K_r

$$P_c = K_r \times Q \tag{4.47}$$

Values of between 0.85 and 1 have been determined for K_r. For temperature calculations, a value of 1 is usually assigned to K_r. Figure 4.24 shows the heat generation and dissipation for a single cutting edge. The converted energy is discharged from the contact zone in a number of ways:

– part of the energy flows into the workpiece;
– part of the energy is discharged by the chips and the coolant;
– the remaining energy flows into the grinding wheel.

The percentage distribution of thermal energy by these routes depends on various factors. A number of authors have come to different conclusions as a result of their studies. Lee (**78**) has observed that about one

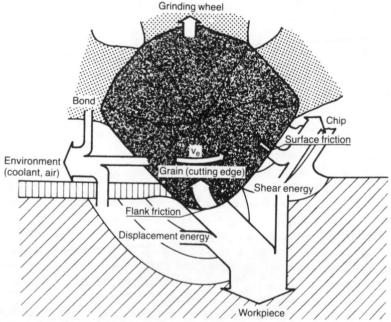

Fig. 4.24 Source of heat and distribution during engagement of the cutting edge (8)

third of the total heat flows into the workpiece, with about half being discharged with the coolant. However, Malkin (79) asserts that 60–80 percent of the heat flows into the workpiece. Brandin (10) claims that 20 percent of the heat flows into the chips and 80 percent into the workpiece. Choi (73) has calculated that, in the case of CBN wheels, 73 percent of the heat from the contact zone flows into the wheel and 27 percent into the workpiece. According to Werner (80), the proportion of heat flowing into the workpiece lies between 30 and 95 percent. It is not possible to assume that there is a fixed distribution of heat between the workpiece, the coolant and the grinding wheel, as there are many factors which can affect the flow of heat during the grinding process.

To obtain more accurate values for the flow of heat into the workpiece, other authors (81)(82) have insulated the workpiece and discharged the grinding heat by means of water flowing through and out of the workpiece. With this method they have been able to determine the quantity of heat by calorimetry.

As it is not possible to measure the temperature in the contact zone by direct methods, and the indirect methods require complicated and time-consuming measuring techniques and preparation of the test piece, many workers have sought to adopt a functional relationship to determine temperature as a function of machine settings and process characteristics.

Werner (83) developed the following temperature model

$$\vartheta = \vartheta_0 \left\{ \frac{K_1'}{z} \times K_2^{\varepsilon_3 \beta K} \times v_c^{2(1-\varepsilon_3)} \times \left(\frac{1}{v_w} \right)^{2\{1-\varepsilon_3(1-\beta K)\}} \right.$$
$$\left. \times d_{se}^{1-\varepsilon_3(1-\beta K)} \times a_e^{\varepsilon_3(1-K)} \right\} \tag{4.48}$$

where

ϑ_0 = starting temperature in °C

K_1, K_2, β = factors dependent on the thermal properties of the material, the coolant system and the wheel grit

K = exponential coefficient, dependent, *inter alia*, on the depth of the cut and the workpiece material

ε_3 = exponential coefficient of the equation defining the cutting force, dependent on the number of cutting edges, the workpiece material and the bond.

In this model, the individual machine settings have to be determined by experiment for each new grinding problem. The application of this model is tied to conditions which, for the most part, limit its use to reciprocating grinding (with a small depth of cut and contact length). It is assumed that $v_w > 50$ mm/s (**83**).

Takazawa (**76**) has developed the following function for temperature distribution in the workpiece

$$\vartheta_z = \frac{2q_w \times a_\vartheta}{\pi \times \lambda \times v_w} \times 3.1 \times L^{0.53} \times e^{(-0.69L^{-0.37}z)} \qquad (4.49)$$

This function can be used to calculate the local temperature beneath the newly created surface. Where $z = 0$, the function for the temperature at the newly created surface is

$$\vartheta = \frac{2q_w \times a_\vartheta}{\pi \times \lambda \times v_w} \times 3.1 \times \left(\frac{l_k \times v_w}{4a_\vartheta}\right)^{0.53} \qquad (4.50)$$

Takazawa's model has been adopted by other authors, either unchanged or with slight modifications (**75**)(**84**)(**85**).

There are other theoretical approaches to the measurement of temperature during grinding (**86**). However, in view of the many different factors affecting the generation of heat in the contact zone, the models proposed to date do not apply universally. Lowin (**9**) has also come to the conclusion that as yet there is no universal model for accurately calculating temperature.

As a rule, the analytical models incorporate constant characteristic quantities such as thermal conductivity and temperature conductivity, which are, however, themselves temperature-dependent, and vary depending on the structure of a material (**87**). These interrelationships are not normally taken into consideration in the analytical calculations, although they can lead to deviations from the temperatures which actually prevail. To date, neither theoretical nor empirical temperature measurements have been carried out for high-efficiency deep grinding with CBN wheels running at substantially higher wheel speeds.

5

Description of the test equipment

5.1 HIGH-EFFICIENCY SURFACE GRINDING MACHINE

A Gühring Automation FS 126 CNC surface grinder was used for the tests into high-efficiency deep grinding (Fig. 5.1).

This machine, specially fitted for research purposes, is characterised by a high spindle output (50 kW) and high running speed (9000 min^{-1}, infinitely variable). Wheel speeds of up to 185 m/s can be achieved with grinding wheels with a maximum diameter d_s of 400 mm.

The grinding spindle is belt-driven by a DC motor with a transmission ratio of 1 : 1.5. The spindle runs on anti-friction bearings and is characterised by great rigidity and true running accuracy (1 μm).

The guides of the machine's three translatory axes are supported by anti-friction bearings and are driven, via bearing-mounted spindles, by DC servo-motors. The minimum controllable travel is 1 μm. Table speed is adjustable in the range 0.02–15 m/min.

Fig. 5.1 Gühring Automation FS 126 surface grinder

Proper balancing of the grinding spindle system is particularly important at these high wheel speeds. This function is performed automatically by an HBA 3001 type hydrocompensator manufactured by Dittel. The hydrocompensator can detect any momentary imbalance in the whole spindle system and rectify any error by injecting coolant into pockets in the face of the flange. This principle works well at speeds of up to 5000 r/min. At higher speeds, however, the injected liquid fails to flow fully into the pockets. In such cases, the system must be balanced beforehand by conventional techniques.

The protective cover was reinforced so as to protect the operator and the machine in the event of the grinding wheel bursting. However, in order to allow the grinding process to be observed and the coolant supply to be monitored, the front gate of the steel cover was replaced with thick armoured glass.

The high wheel speed used in high-efficiency deep grinding gives rise to a heavy oil mist in the grinding area. To forestall any harm to the health of the operator, the machine was fitted with an extractor and an electrostatic air-filter. This system is mounted directly on the grinding machine. Approximately 4000 m^3/h of air are extracted from the grinding area and filtered.

For the purpose of the tests, the machine was also fitted with:

- a powerful coolant system (built by the author);
- a dressing system using the swing-step technique (developed in cooperation with the company KW-Abrichttechnik),
- a sharpening device (built by the author),
- an adaptive control unit to reduce vibration (lent by Diamant Boart Deutschland GmbH, Haan).

The coolant system cleans and delivers the coolant at a rate of up to 500 l/min at a pressure adjustable up to 20 bar. The coolant system, the coolant supply and the cleaning of the wheel are considered in detail in section 4.4.

One of the characteristics of the swing-step dressing system is low vibration during profiling, the effect of which is to reduce any deviation on the profile. The specific characteristics and the kinematics of the swing-step process are described in section 4.5.

Details of the high-performance sharpening system used to cut back the bond on dressable CBN wheels are also given in section 4.5.

In collaboration with Leuven University in Belgium, Diamant Boart have developed an adaptive control unit which enables speed-dependent vibrations to be reduced automatically by moving out of the critical wheel speed range. In the event of unacceptably high vibrations, which occur at certain speeds, the unit switches down to preprogrammed, lower speeds. Once the amplitude of the vibrations has returned to the normal range, the wheel speed is stepped up again to the original programmed value. The unit can prevent the chatter marks which are sometimes caused at critical speeds. It has been successful when machining long workpieces and at low workpiece speeds, i.e., it produces good results in grinding operations which, as far as workpiece speed is concerned, correspond to the conditions encountered in creep-feed grinding. However, it is less suitable for HEDG operations, which require only a few seconds, as it takes time to measure the vibrations and adjust the rotational speed of the grinding wheel.

5.2 MEASURING EQUIPMENT

A Kistler three-component dynamometer in the form of a measuring plate was used to measure the grinding forces. In this system, the three orthogonal components of a force are detected by piezoelectric force sensors. For each of the three force components, a proportional electric charge is generated in the dynamometer, which is converted into an analog DC voltage in the charge amplifier. An A/D converter, which is built into a 16-bit measuring computer, converts this into corresponding digital values which, after further processing by a software program, are produced in the form of tables or diagrams.

A separate charge amplifier is used for each force component. The low-pass filter connected downstream from the charge-amplifier input stage suppresses the high-frequency interference vibrations occurring while the measurement is being taken. The block diagram of this arrangement is shown in Fig. 5.2.

A Talysurf 5 stylus instrument, made by Rank Taylor-Hobson, was used to determine the surface characteristics.

A tracing device developed by the Production Engineering Department was also used to determine the topography of the grinding wheels (**88**). The device consists of a very true-running, high-precision spindle to receive the grinding wheel, a small wheel driven by a reducing gear of a stepping motor, and the tracing head of a (Talysurf 5) measuring apparatus, together with its associated evaluation unit (Fig. 5.3).

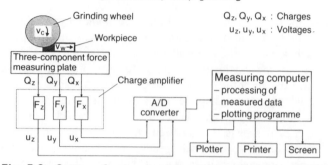

Fig. 5.2 System for measuring grinding forces in HEDG

The tracing head makes an axial trace of the grinding wheel, after which the small wheel turns the grinding wheel through a very small angle, so that the line of the next trace lies at a distance of about 10 μm from the first trace. This process is repeated until a representative area of the abrasive layer has been covered. A program processes the measured

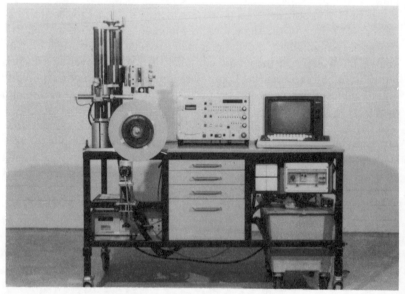

Fig. 5.3 Tracing device for determining the topography of the grinding wheel

data and the topography of the area examined is plotted out in enlarged form, enabling changes to be detected and the condition of the grains and the bond to be checked.

The electrical output of the spindle was also measured using a G 72-PS wattmeter, made by Valenite–Modco. An instrument transformer measures the instantaneous voltage and current at the spindle motor and multiplies the two values. The product of the calculation is shown as the active power (spindle output) on a display built into the measuring unit, and is inputted into the process computer via an A/D transformer for further processing.

The residual stresses produced in the surface zones of the workpiece were measured by X-ray by the Institut für Werkstofftechnik (IWT) using a goniometer. The measuring technique rests on detecting the relative change in the distance between atoms caused by the machining process. The state of residual stress in the surface zone can then be established by applying the relationships which are well known from the theory of elasticity. The results of the measurements, and the assessment of the results, are presented in section 6.6.

The grinding temperatures in the vicinity of the contact zone between the workpiece and the grinding wheel were measured with a number of sheathed thermocouples. The measuring method is described in detail in the next section.

5.3 APPARATUS AND METHOD FOR MEASURING TEMPERATURE

Many different methods have been used to measure temperature during grinding, for example:

- photoelectric cells;
- thermal resistors;
- heat-sensitive paints;
- single-wire thermocouples;
- two-wire thermocouples (sheathed and unsheathed);
- infrared temperature measurements.

Single and two-wire thermocouples are the most commonly used methods for measuring grinding temperature, enabling fairly accurate measurements to be made in wide temperature ranges. They can be placed in the contact zone, or just below, by the provision of holes or other means.

In order to reduce the size of the thermocouples, and consequently to reduce further their response time, Choi (**73**) has sought, on the basis of two-wire thermocouples, to produce thin thermocouples by means of vapour deposition on various substrates. Although so far he has been unable to achieve satisfactory results, the further development of thermocouples of this kind may lead to the more accurate measurement of temperatures in the contact zone.

It is virtually impossible to measure temperature in the contact layer or while a cutting edge is in engagement with the workpiece. Lowin (**9**) has shown that the methods used are not suitable for accurately measuring the grinding temperature in the contact zone. However, attempts are being made to determine the average surface-zone temperature by the interpolation and extrapolation of measurements taken at various points below the surface layer.

The response time of the thermocouples is critical to determining the rapidly changing grinding temperature. The response times of sheathed thermocouples (which were also used in these studies) are shown in Fig. 5.4. The measurements were taken in boiling water and give 9/10-value times for two types of thermocouple and various diameters.

In this study the temperatures were measured using sheathed thermocouples with an outside diameter of 0.5 mm, placed in and beneath the surface zone, and at various points in the contact zone. They consist of NiCr/Ni wires clad in a protective Inconel sheath. The wires are insulated up to the welding point. According to DIN 43710 (1984 draft) *Basic values for thermoelectric voltage in mV for Fe–CuNi*, and information supplied by the manufacturers (**90**), they are suitable for measuring temperatures up to about 1000°C. CrNi/Cr thermocouples are known as type K, and possess good linearity over the whole measurement range.

As thermocouples require a reference temperature, a measuring card with an accurate temperature detector was used. The measuring card measures the instantaneous room temperature and feeds it into the process computer as the reference temperature. Fifteen thermocouples can be connected to the measuring card simultaneously. The measurement signals from the thermocouples are fed to the measuring card and amplified in the process computer together with the signals from the reference position. A suitable software program processes the data and displays it on a monitor and plotter. Calibration tests in icy water,

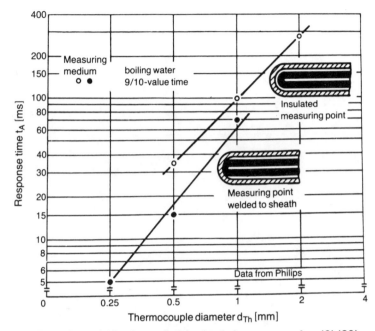

Fig. 5.4 Response time of sheathed thermocouples (9)(89)

boiling water and at high temperatures in an electric furnace demonstrated that the expected response times were correct and that the temperature could indeed be measured accurately.

In order to measure the temperature, five holes were drilled just below the surface of the workpiece and the thermocouples were inserted. Figure 5.5 shows the distribution of the measuring points at distances of 0.1–1 mm below the newly produced workpiece surface. The tips of the thermocouples were made to penetrate into the workpiece to a depth of 5 mm, corresponding exactly to the centre-line of the grinding wheel. A device was used to thrust them into the hole, where they were clamped in position. A conductive paste was introduced into the hole to improve the contact between the temperature sensor and the workpiece.

Figure 5.6 shows a workpiece which has been prepared for the temperature to be measured. The workpiece can be used twice, and is designed for a depth of cut of 10 mm. A clamping device is attached to the central borehole to secure the thermocouples.

Figure 5.7 shows the temperatures recorded by the five thermocouples

High efficiency deep grinding

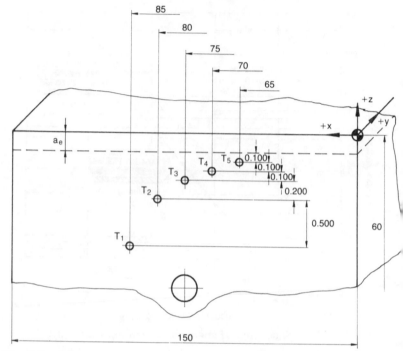

Fig. 5.5 Arrangement of the thermocouples for measuring temperature

positioned at different depths. The workpiece material, 42CrMo4, has a hardness of 34 HRC. The test was carried out with a B252 grit electroplated CBN wheel. The depth of cut a_e was 5 mm and the specific removal rate Q'_w was 30 mm^3/(mm · s).

The temperature/time curves T_1–T_5 are those of the five thermocouples. T_5 is the temperature curve for the thermocouple placed closest to the surface zone. In this case, there is a difference in temperature of more than 100°C between the first and the last temperature sensor. A temperature of about 500°C was measured 0.1 mm beneath the newly produced workpiece surface. The staggering of the maximum temperatures is understandable, as the thermocouples were set 5 mm apart in the grinding direction (as in Fig. 5.5). The temperature sensors had to be offset in relation to each other, as in some cases their diameter was greater than their offset in depth.

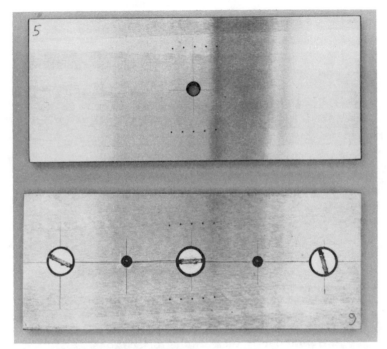

Fig. 5.6 Pre-drilled workpiece for measuring temperature

The average surface-zone temperature at the grinding surface was calculated by extrapolation.

The grinding tests also showed that, with sufficiently long workpieces, the temperature is distributed constantly shortly after the full contact length is reached or shortly before it has ended, provided there is no change in the grinding conditions. It was also observed that, in the case of burnt workpieces, damage to the material extends to the same depth over the entire workpiece, provided that the contact length remains constant.

The temperature is not uniform in the contact zone itself. A number of workers maintain that temperature is at its highest at the centre of the contact length (**20**).

In order to measure the temperature in the contact zone, the first step was to determine the contact length for particular grinding wheel diameters and depths of cut. Five holes were then drilled in five positions

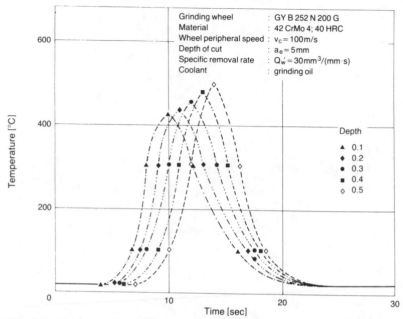

Fig. 5.7 **Temperature/time graph for five thermocouples located at different depths beneath the surface zone**

0.1 mm away from the contact zone (Fig. 5.8). The holes extended to the centreline of the contact width and were 0.5 mm in diameter.

The measurements were made using the same technique as was used

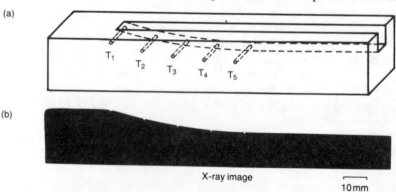

Fig. 5.8 **Temperature measuring points below various points in the contact zone (X-ray photograph)**

to determine the temperature of the surface zone. The grinding operation had to be finished 0.1 mm above the thermocouples and precisely at the preset contact point. The workpieces and holes must be prepared precisely in order to obtain an accurate measurement and to prevent the thermocouples from being damaged by the wheel.

Contact-zone temperatures under various working conditions were determined with this method. The results are described in section 6.5.

6

Test results

6.1 SPINDLE OUTPUT AS A FUNCTION OF GRINDING PARAMETERS

In high-efficiency deep grinding, spindle output is of great importance for the design of the grinding machine and for determining the grinding parameters. It is a characteristic of the method that HEDG requires a spindle output many times greater than that needed for conventional creep-feed grinding.

Spindle output or grinding power depend on a number of factors, the most important of which are:

- grinding parameters: wheel speed, depth of cut, workpiece speed;
- grinding wheel: wheel diameter, wheel width, grit type, grit size, condition of the grit (sharpness), grit protrusion (dressing and sharpening conditions), concentration, bond, structure;
- coolant: viscosity, pressure, quantity, temperature, shape and number of delivery nozzles, cleaning system;
- workpiece: material, geometry.

The effect of these many factors on grinding performance will not be considered in detail in this work. For high-efficiency deep grinding, the power requirement can be broken down into the following two components:

- overcoming the braking effect in the grinding profile with a running grinding wheel and the use of coolant;
- the (net) cutting power.

Figure 6.1 shows spindle output as a function of wheel speed. The test consisted of grinding a 6.5 mm deep involute tooth system with a CBN wheel using oil as coolant.

If no coolant is used, the no-load output P_L increases only slightly as wheel speed increases, whereas the spindle output is markedly higher when the coolant system is engaged, due to the braking effect in the premachined groove, and rises from about 12 to 25 kW when the wheel speed is increased from 120 to 180 m/s. The total output P_{ges} also increases with the same rise in the wheel speed. The difference between

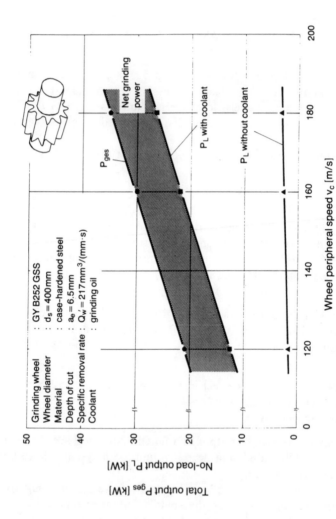

Fig. 6.1 Spindle output as a function of wheel speed

total output and no-load output is virtually constant, and this value represents the metal cutting output (**12**)(**34**).

Figure 6.2 shows spindle output in relation to workpiece speed in the surface grinding of cast iron with a CBN wheel, before and after touch-dressing. The electroplated CBN wheel (B252 grit) was touch-dressed 30 μm (from the radius) using a diamond roll, up-cutting with a low axial feed (1 mm/s) in 6 passes with a 5 μm depth of cut in each pass.

The spindle output with the touch-dressed wheel is approximately 30 percent higher. The increase in the power drawn is due to the flattening (reduced sharpness) of the CBN grains and the reduction in grit protrusion and chip space.

Despite the large amount of power needed to overcome the braking effect in HEDG, the energy required to remove 1 mm^3 of stock is much lower than in conventional creep-feed grinding. Figure 6.3 shows the specific power consumption for grinding similar grooves with conventional creep-feed grinding HEDG. While the braking and no-load outputs are much lower in conventional grinding than in HEDG, owing to the lower wheel speed and the use of emulsion as coolant, the specific energy

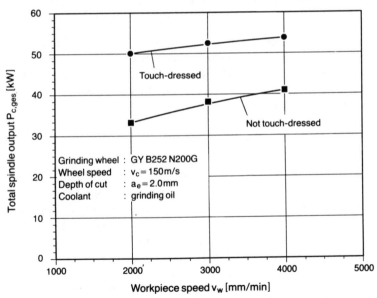

Fig. 6.2 Spindle output for a CBN wheel before and after touch-dressing

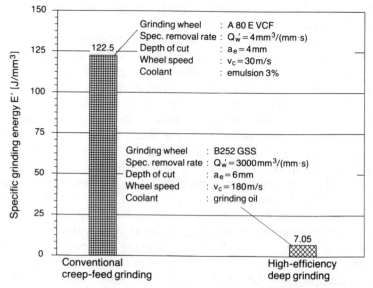

Fig. 6.3 A comparison between specific energy requirements in conventional creep-feed grinding and HEDG

requirement in HEDG is, however, much lower, amounting to only about 6 percent of that required in conventional creep-feed grinding. This reduced energy requirement is a further advantage of high-efficiency deep grinding and is one of the reasons why the workpiece does not suffer thermal damage.

6.2 GRINDING FORCES AS A FUNCTION OF GRINDING PARAMETERS

Total grinding force is the sum of all the forces acting on the cutting edges which are engaged at any one moment. It consists vectorially of the normal and tangential force components. In physico-mechanical terms, the product of the tangential force component and wheel speed corresponds to the grinding power **(8)–(10)(14)(15)**. As the peripheral speed of the wheel increases, the grinding forces decrease. As the number of cutting edges engaging the workpiece decreases, the average chip thickness must also become smaller. A number of authors have described and confirmed this behaviour at low and medium wheel speeds **(8)–(11)(14)(15)(91)**.

The grindability coefficient, ε, is a material-specific value which has a decisive influence on the grinding force. Materials with poor thermo-mechanical grindability have a low coefficient (around 0.5), which results in very high total grinding forces, caused primarily by frictional processes. On the other hand, materials which are easier to grind have a grindability coefficient approaching the value of 1. Figure 6.4 is a qualitative presentation of grinding force as a function of wheel speed for two different groups of materials (4). The first group, with a predominant chip-forming force component, which is characterised by high values of ε exhibits a marked decline in grinding force as wheel speed v_c increases. These materials generally possess good grinding properties. The second group, with a predominant frictional power component, and which is characterised by low values of ε exhibits only a slight drop in grinding forces as wheel speed v_c increases. Materials of this type are difficult to grind and high grinding forces are generated at high wheel speeds (4). This means that the intensity of the decrease in grinding force as the

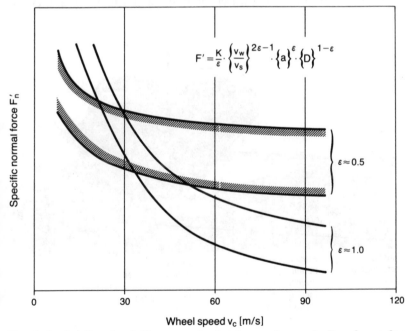

$$F' = \frac{K}{\varepsilon} \cdot \left\{ \frac{v_w}{v_s} \right\}^{2\varepsilon - 1} \cdot \left\{ a \right\}^{\varepsilon} \cdot \left\{ D \right\}^{1 - \varepsilon}$$

$\varepsilon \approx 0.5$

$\varepsilon \approx 1.0$

Specific normal force F_n'

Wheel speed v_c [m/s]

Fig. 6.4   Qualitative influence of wheel speed on grinding force for different types of material (4)

peripheral speed of the wheel increases is a function of the workpiece material.

Ernst (**92**) and Daude (**72**) have explained the decrease in grinding force as the wheel speed increases as being due to more favourable kinematic conditions. Gühring (**11**) confirms this beneficial effect of more favourable kinematic conditions, and also demonstrates that reduced mechanical strength at elevated temperatures has a critical influence on the reduction in grinding force.

However, the reasons given for the reduction in grinding forces do not all apply in high-efficiency deep grinding. Increasing the wheel speed under otherwise constant conditions reduces the average chip thickness and also the number of cutting edges engaging the workpiece, which in turn results in a reduction in the grinding forces. However, in high-speed grinding the cutting forces are lower, even if the grinding operation is performed with unchanged chip thickness.

Prins (**93**) has established with single-point tests that the tangential force drops with increasing wheel speed and a constant depth of cut. In

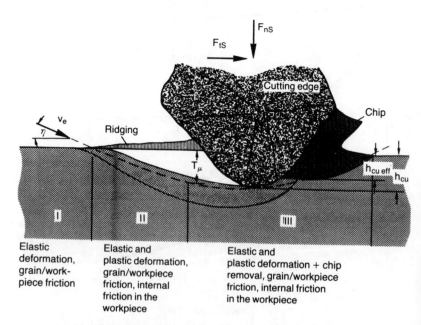

Fig. 6.5 Cutting edge engagement in grinding (8)

this case the explanation does not lie in any reduction in the cross-section of the chip, as the depth of cut in the groove remains constant. The reason for the decrease in the force which accompanies an increase in wheel speed must be sought in the chip formation process. That is to say, the reason lies in the reduced energy, and hence force, needed for elastic and plastic deformation to occur when the cutting edge engages the workpiece. Higher wheel speeds also result in a less ductile behaviour of the material as it is deformed, and so reduce the force and energy input required for the grinding operation.

According to König (8) and Stefens (75), as the cutting edge engages the workpiece, the initial mechanism is that of elastic deformation. As the edge penetrates further, elastic and plastic deformation occur, followed ultimately by the formation of the chip. Through all these stages elastic deformation remains an element in the cutting process. Figure 6.5 shows the deformation which takes place during the various stages as the cutting-edge engages the workpiece during grinding (8). The energy required for elastic and plastic deformation is considerable. In HEDG, because of the faster wheel speed and the high temperature of the contact

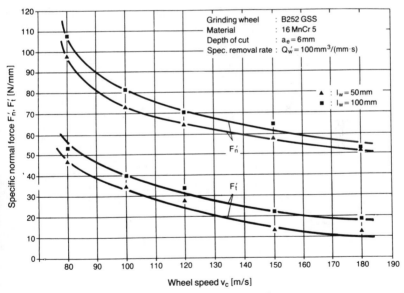

Fig. 6.6 Normal and tangential forces as a function of wheel speed for different grinding lengths

layer (not the workpiece surface temperature), chip removal occurs with a lower degree of deformation, resulting in less force and energy being required.

Figure 6.6 shows grinding forces as a function of wheel speed.

Workpieces measuring 50 and 100 mm in length were ground in the up-cutting mode at a constant depth of cut a_e of 6 mm. For the purpose of these investigations, the specific removal rate Q'_w was kept constant at 100 mm³/(mm · s).

The reduction in grinding forces as the wheel speed increases is apparent. It is due to the diminishing chip thickness and the reduction in the dynamic number of cutting edges at high wheel speeds. Because of the lower grinding forces, the increased wheel speed leads to reduced deformation of the workpiece.

Figures 6.7 and 6.8 show grinding forces as a function of specific removal rate. In these investigations the specific removal rate Q'_w was increased to values of up to 3000 mm³/(mm · s). Grooves with a high-quality surface finish were ground at wheel speeds v_c of 180 m/s using two grinding wheels, each containing a different size grit (B252 and B151). At a depth of cut a_e of 6 mm, it was possible to grind up to a Q'_w value of only 1500 mm³/(mm · s), as under these conditions the

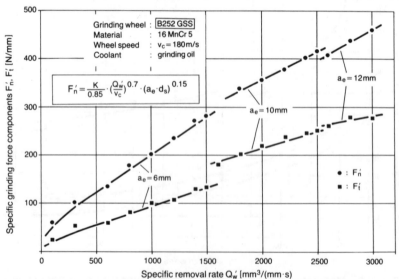

Fig. 6.7 Grinding forces for an electroplated B252 CBN wheel

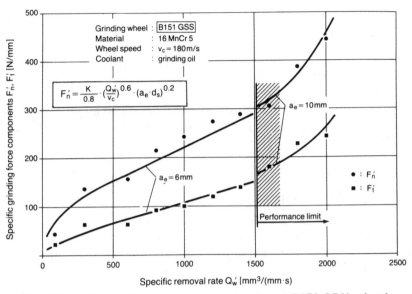

Fig. 6.8   Grinding forces for an electroplated B151 CBN wheel

maximum feed rate of the machine (15 000 mm/min) was reached. In
order to increase further the specific removal rate, depth of cut was
increased to 10 and 12 mm. The force curves in Figs 6.7 and 6.8 consist
of three ranges, characterised by the specific conditions below:

Range 1:	$Q'_w = 0$–1500 mm³/(mm · s)	$a_e = 6$ mm
Range 2:	$Q'_w = 1500$–2500 mm³/(mm · s)	$a_e = 10$ mm
Range 3:	$Q'_w \geq 2500$ mm³/(mm · s)	$a_e = 12$ mm
	(B252 wheel only)	

The transition from range 1 to range 2 is characterised by an abrupt
increase in grinding forces. This is due to the increase in the contact
length caused by increasing the depth of cut a_e from 6 to 10 mm. In
range 3, increasing the depth of cut from 10 to 12 mm does not result in
any increase in the normal force.

At a depth of cut of 12 mm it was possible to raise the specific removal
rate Q'_w with the B252 wheel to a maximum of 3000 mm³/(mm · s). This
value does not, however, represent the performance limit for the process
and for this grinding wheel, but the tests could not be taken any further
as the maximum feed rate possible with this machine had been reached.

With the finer-grained B151 wheel the removal rate limit was reached at a Q'_w value of 1500 mm³/(mm · s). At this rate, the chip spaces could not accommodate any more material. A further rise in the specific removal rate to 2000 mm³/(mm · s) results in a disproportionate increase in grinding forces. The rate could be increased by stepping up the wheel speed.

The contact length has a critical effect on grinding forces and the progress of the operation. It can be varied by adjusting the following parameters:

- depth of cut;
- grinding wheel diameter;
- length of the workpiece.

The length of the workpiece affects the contact length l_k only if it is smaller than the contact length which can be obtained with a given grinding wheel diameter and depth of cut. The functional relationship for the contact length can be expressed as follows **(8)(9)(13)**

$$l_k = \sqrt{(a_e \times d_s)} \tag{6.1}$$

In the case of short workpieces ($l_w < l_k$), the grinding forces F' (relative to the width of the grinding wheel) are lower, but the area-related grinding forces are higher. Figure 6.9 shows the area-related normal force F''_n as a function of workpiece length l_w for various depths of cut. The figure shows two parameters, workpiece length and depth of cut, which are critical in determining the contact length.

With workpieces of sufficient length, e.g., where l_w or $l_s > l_k$, the forces are determined only by the depth of cut a_e. Smaller depths of cut result in greater area-related grinding forces in the normal direction than larger depths of cut. The reason for this is the shorter contact length with smaller depths of cut. With a depth of cut of 12 mm and a wheel diameter d_s of 400 mm, the equation (6.1) gives a contact length l_k of 69 mm.

As can be seen from the figure, force F''_n increases with diminishing workpiece length down to a workpiece length of 69 mm. Where the depths of cut are 9 mm, 6 mm, and 3 mm, the resulting contact lengths l_k are 60 mm, 49 mm, and 35 mm, respectively.

With short workpieces, and also when the wheel starts to engage the workpiece, the full contact length is not attained. The area-related grinding forces therefore remain high. As the length of engagement increases, the value F''_n falls steadily until the full contact length is reached. With

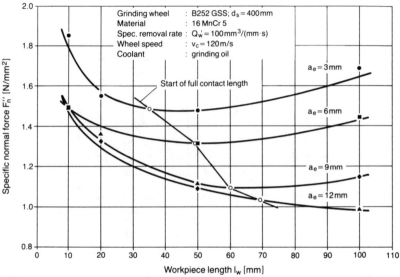

Fig. 6.9   Area-related grinding force F_n'' as a function of workpiece length for various depths of cut

the depth of cut values of 3 mm, 6 mm, and 9 mm, the area-related normal forces continue to rise. This may be explained by the increasing friction between the grinding wheel and the workpiece. This means that, when the full contact length is attained, only half the grinding wheel has penetrated the workpiece. As the process continues, the contact length remains constant, but the other half of the wheel causes the grinding forces to increase. This frictional effect brings about increased grinding forces only until the grinding length is about twice the contact length. Thereafter the grinding forces remain constant, provided that l_k remains unchanged.

It is not only the contact length, but also the width of contact, or the grinding width, which is important in HEDG. In addition, the (lateral) shape of the contact area can also have a considerable influence. These problems are discussed below in the context of a grinding operation (**33**).

The task involved grinding a groove with a depth of cut a_e of 25 mm, a width at the base of 1.0–1.5 mm, and a flank angle of 2 degrees in a hot-work tool steel (45NiCr6; 35 HRC) (Fig. 6.10). Groove profiles of this kind are frequently found in injection-moulding tools, for example.

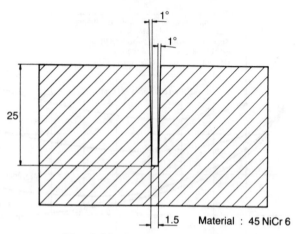

Fig. 6.10 The groove profile

A further requirement was that, if possible, the groove should be produced with the desired surface finish in a single pass. A high-quality finish on the flanks of the groove was necessary in order to ensure that the plastic injection mouldings to be produced subsequently could be released cleanly and easily from the mould. An average roughness value R_a of about 1.0 μm was required. In addition, the sides of the groove had to be free from any thermal damage (burns, cracks).

Compared with normal, rectangular grooves, the main problem in producing profiles of this kind, with shallow flank angles, lies in the greatly increased contact area between the grinding wheel and the workpiece. This causes high cutting and frictional forces, which increase the risk of thermal damage to the workpiece. Special attention must therefore be paid to the proper delivery of the coolant. The specification and design of the grinding wheel – and especially the selection of the correct grit type – are critical to achieving good results, particularly when machining unhardened steels.

Three different electroplated CBN wheels were used to carry out this machining operation. The flanks of the first two were fully coated with B151 grit. The second wheel was provided with radial slots. Neither type of wheel produced satisfactory results, but they did illustrate the requirements which needed to be met for a wheel to perform the task in question.

The most suitable tool proved to be a CBN wheel electroplated with B251 grit, not fully coated on the sides (Fig. 4.2). This is a novel design. The experience gained with the CBN wheels used previously led to the following design modifications, which were further optimised in practical operation.

(1) Prestressed-steel wheel body.
(2) Coarser grit – B251 instead of B151.
(3) Partial coating of the grinding wheel flanks. This serves to reduce the high lateral friction between the grinding wheel and the work-piece, and at the same time improves the delivery of coolant to the lateral contact zones.

The degree to which the wheels were coated was governed by the consideration that the abrasive coating on the flanks has to remove the same specific volume of material as the coating on the periphery of the wheel. With a flank angle of 2 degrees, the volume of stock to be removed by the flank of the wheel only amounts to about 4 percent of the volume of metal removed at the peripheral surface. In order to bring about a uniform rate of metal removal in all areas of the wheel, the abrasive coating on the flanks was reduced accordingly. The periphery of the wheel and part of the flanks (up to a height of 1.5 mm) are fully coated, while both flanks are provided with 18 coated segments (measuring 10 mm in width and 30 mm in height). These segments are positioned exactly opposite one another, so as to prevent any alternating transverse forces from occurring during grinding.

These measures resulted in a high-performance wheel of novel design exhibiting the following features:

– a prestressed wheel core, resulting in a grinding wheel of increased stability;
– the use of a coarser grit for machining the relatively soft steel, resulting in reduced cutting forces, higher removal rates, and very good roughness values;
– partial coating of the wheel flanks, resulting in reduced friction and lower cutting forces, but also lower specific removal rates.

The optimisation of the grinding wheel and the use of suitable coolant and cleaning nozzles resulted in acceptable grinding forces.

Figure 6.11 shows the grinding force components for depths of cut a_e of up to 20 mm at various wheel speeds (v_c = 120 m/s and 180 m/s). At a

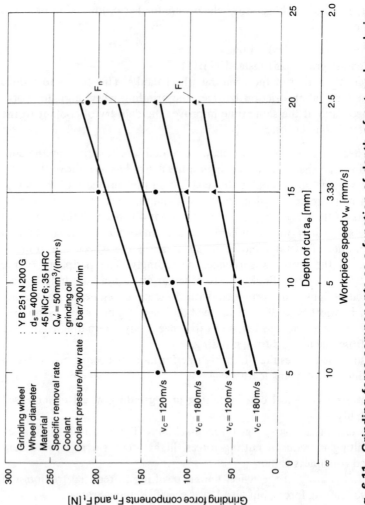

Fig. 6.11 Grinding force components as a function of depth of cut and workpiece speed at a constant specific removal rate

constant specific removal rate, an increase in the depth of cut and a corresponding reduction in workpiece speed cause the grinding forces to increase. This is due to the increasing contact length and the consequent increase in the number of kinematic cutting edges.

6.3   WORKPIECE ROUGHNESS AS A FUNCTION OF GRINDING PARAMETERS

One of the most important results of a grinding operation is the surface roughness generated on the finished workpiece.

Figure 6.12 shows the relationship between peak-to-valley height R_z and wheel speed for depths of cut a_e of 5 mm, 10 mm, and 15 mm. Figure 6.13 shows the same relationship for the centreline average height R_a. 8 mm wide electroplated CBN wheels with different grit sizes (B252 medium and B427 coarse) were used in the tests.

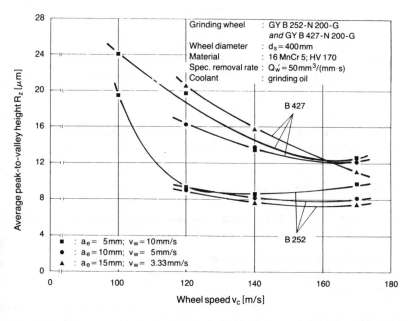

Fig. 6.12   Average peak-to-valley height R_z as a function of wheel speed for various grit sizes and depths of cut

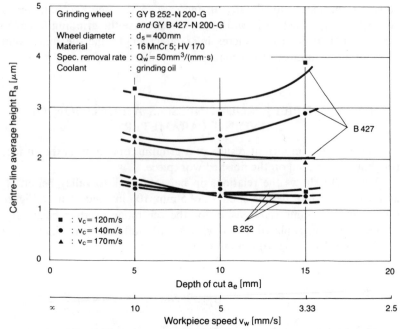

Fig. 6.13 **Centreline average height R_a as a function of depth of cut a_e and workpiece speed v_w for various grit sizes and wheel speeds**

The following conclusions can be drawn with regard to the average peak-to-valley height R_z and the arithmetical centreline average height R_a.

(a) The roughness of the workpiece decreases steadily as the wheel speed v_c rises (Fig. 6.12).

(b) A coarser grit (B427) results in greater roughness values (Figs 6.12 and 6.13).

These observations are in agreement with analytical findings (**94**). The roughness values established in the test ($R_z = 8$–24 μm) are in the expected range.

The roughness values obtained in high-efficiency deep grinding using CBN and aluminium oxide wheels are in line with those obtained by conventional grinding. The increase in roughness produced by increasing

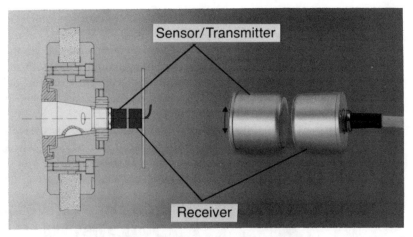

Fig. 6.14 Arrangement of the sensor/transmitter on the shaft of the grinding wheel spindle (after Dittel)

the workpiece speed in HEDG is partly offset by the large depth of cut and high wheel speed. With a specific removal rate about a hundred times greater than that achieved in conventional grinding, high-efficiency deep grinding can, nevertheless, produce similar roughness values.

If special techniques are used, such as touch-dressing the wheel with a diamond tool, substantially lower roughness values ($R_z = 2$–4 μm) can be achieved with high-efficiency profile grinding.

Despite the greater sensitivity of the acoustic sensor and its electronics, compared with force tranducers, identification of the start of the dressing cut is still dependent on the accuracy and transmission capacity of the wheel spindle bearings. The sensor is generally attached to the grinding machine's headstock or the housing of the dressing apparatus. The signal to indicate that the wheel and the touch-dressing tool are in contact must be transmitted to the sensor via the bearings of the spindle or the dressing apparatus. Insufficient rigidity of the bearing, or the presence of a film of air and oil, can cause the signals to be lost or result in a loss in transmission. The start of the cut will thus be identified wrongly. This problem was discussed with the Dittel company and, after much discussion, a solution was found. An acoustic sensor, coupled with a small transmitter, was developed. The sensor and the transmitter can be attached to the grinding wheel spindle shaft or directly to the dressing apparatus and rotate with the shaft. A small receiver at a slight distance

(1 mm) from the transmitter/sensor provides for contactless transmission of the signal (Fig. 6.14).

The signals to denote contact no longer need to be transmitted via the bearings. They pass the fixed connections of the grinding wheel with the shaft or of the touch-dressing tool with its shaft and reach the sensor/transmitter. The transmitter sends the signal to the receiver, which, in turn, transmits the signal to the evaluating electronics (**95**)(**96**).

Initial studies with the transmitter/sensor system have produced excellent results. A workpiece was ground with an electroplated CBN wheel (B151 grit size) and the roughness was measured (Fig. 6.15(a)). The workpiece was then ground again using the same parameters but after the

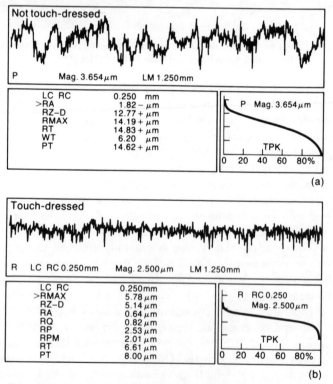

Total material removal (radial) by touch-dressing $a_{ed,ges} = 10\,\mu m$

Fig. 6.15 Workpiece surface finish before and after touch-dressing (radial removal 10 μm)

wheel had been touch-dressed by 10 μm. The surface roughness values were reduced and a finer surface, as is generally required, was obtained. The R_a value was improved by almost the same amount as that removed by touch-dressing (Fig. 6.15(b)). An optimum touch-dressing result depends primarily on the amount of wheel surface removed. An exact amount can, obviously, only be achieved if the start of the cut is identified accurately.

A radial amount of 5–7 percent of the grit particle diameter is recommended as the optimum amount for touch-dressing. The 'shaving' of larger amounts from the wheel does not normally result in a reduction in workpiece surface roughness, but in higher grinding forces and grinding power. Excessive touch-dressing will of course also reduce the life of the wheel **(95)(96)**.

Figure 6.16 shows a comparison of the roughness values obtained with an electroplated CBN wheel before and after touch-dressing. The wheel was touch-dressed with a diamond dressing roll by an amount of 30 μm

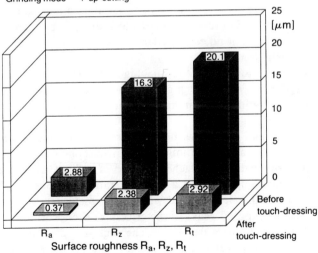

Grinding wheel : GY B 252 N 200 G
Wheel speed : $v_c = 150$ m/s
Workpiece speed : $v_w = 50$ mm/s
Depth of cut : $a_e = 2$ mm
Coolant : grinding oil
Grinding mode : up-cutting

Fig. 6.16 Comparison between the roughness values produced by a CBN wheel before and after touch-dressing

in six increments each of 5 μm. The touch-dressed grinding wheel produces a surface finish with a value approximately 85 percent less than that produced without touch-dressing.

Touch-dressing removes the exposed peaks of the abrasive grains, thus causing a greater number of grains to engage the workpiece. This in turn gives rise to a more uniform cutting process, with reduced chip thickness and hence an improved surface finish.

The marked improvement in workpiece roughness when touch-dressed wheels are used enables coarser CBN grains to be employed. Furthermore, with regard to surface finish, a touch-dressed CBN wheel is more competitive in relation to conventional grinding wheels.

Because of the greater forces involved and the short machining time, a touch-dressed grinding wheel is likely to result not only in reduced surface roughness, but also in greater, and usually desirable, compressive residual stress (see section 6.6).

6.4 WEAR AND GRINDING RATIO AS A FUNCTION OF GRINDING PARAMETERS

A comparison is made below between radial grinding wheel wear relative to specific volume of metal removed for an aluminium oxide wheel and two CBN wheels. Figure 6.17 shows the radial wear of an aluminium oxide wheel used to grind case-hardened steel as a function of the specific volume of metal removed.

The relatively low wear and the high grinding ratio (conventional wheels) of $G = 78$ show that high-efficiency deep grinding is more economical than traditional grinding techniques even when conventional wheels are used.

Figures 6.18 and 6.19 show the radial wear for a (single-layer) electroplated CBN wheel and for a (multi-layer) metal-bond CBN wheel.

The electroplated wheel reached the end of its life after approximately 2.8×10^6 mm³/mm, attaining a grinding ratio G of 35070. The radial wear pattern of the electroplated wheel is characterised by severe initial wear. Once the exposed peaks of the grits have been worn down, the rate at which the tool wears decreases. The high grit concentration of the single-layer CBN wheel is the reason for the low wear and higher G ratio.

The multi-layer CBN wheel wears somewhat more quickly than the

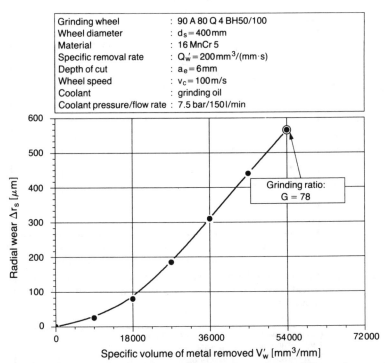

Grinding wheel	: 90 A 80 Q 4 BH50/100
Wheel diameter	: $d_s = 400\,\text{mm}$
Material	: 16 MnCr 5
Specific removal rate	: $Q'_w = 200\,\text{mm}^3/(\text{mm}\cdot\text{s})$
Depth of cut	: $a_e = 6\,\text{mm}$
Wheel speed	: $v_c = 100\,\text{m/s}$
Coolant	: grinding oil
Coolant pressure/flow rate	: 7.5 bar/150 l/min

Grinding ratio:
G = 78

Fig. 6.17 Radial wheel wear as a function of specific volume of metal removed for an aluminium oxide wheel

single-layer wheel. However, with a *G* value of almost 25 000, it has a grinding ratio which cannot be achieved in conventional grinding.

Due to the more favourable cutting process in HEDG compared with conventional reciprocating and creep-feed grinding, the tool wear is very low. At the same time, the grinding ratio (*G* ratio) is well in excess of that normally achieved with conventional grinding wheels. The wear is particularly low when extremely hard CBN abrasive is used. Grinding ratios (*G*) of 20 000 and more can be achieved (the grinding ratio is the ratio of the volume of workpiece material removed relative to the volume of tool wear). It follows that a long tool life can be achieved with a CBN wheel with an abrasive coating just a few millimetres thick. If these wheels are used to their best advantage, it is possible not merely to produce superior surface finishes, but to machine, with less tool wear, more economically than by turning or milling.

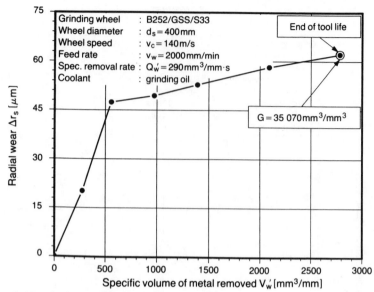

Fig. 6.18 Radial wheel wear as a function of specific volume of metal removed for a single-layer, electroplated CBN grinding wheel

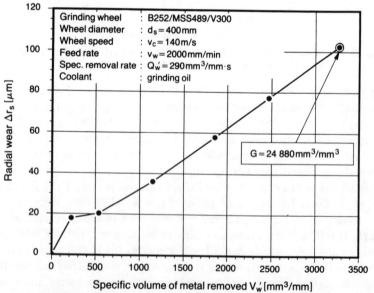

Fig. 6.19 Radial wheel wear as a function of specific volume of metal removed for a metal bond, multi-layer CBN grinding wheel

6.5   TEMPERATURE AS A FUNCTION OF GRINDING PARAMETERS

Figure 6.20 shows the way in which the temperature at the surface of the workpiece tends to change depending on the grinding parameters. It is apparent from the figure (top left) that, in HEDG, the temperature in the surface zone falls as the speed of the workpiece increases. In high-efficiency deep grinding, the increase in workpiece speed means that the heat source (the grinding wheel) has less time to act on the workpiece, which therefore absorbs less heat. The figure (top right) also shows that the surface zone temperature increases with increasing depth of cut. In conventional creep-feed grinding, by contrast, an increase in the depth of cut (with the stock removal rate remaining constant) is accompanied by a fall in the temperature in the surface zone. The rise in temperature in HEDG is due to the increasing contact length, greater friction, and greater energy conversion.

Figure 6.20 also shows surface zone temperature relative to wheel speed for an aluminium oxide wheel and a CBN wheel. The temperature

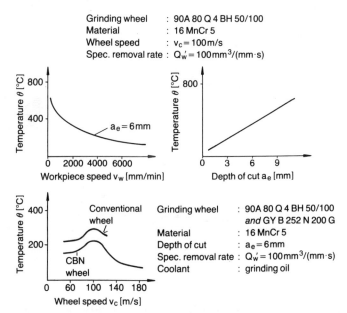

Fig. 6.20   Workpiece surface temperature as a function of grinding parameters

level for the aluminium oxide wheel is higher than for the CBN wheel, owing to the higher thermal conductivity and the smaller negative rake angle of the CBN wheel. The temperature rises continuously up to a wheel speed of approximately 100 m/s. This thermal behaviour is in agreement with the findings of other authors (4)–(6)(8)(9)(11)(13)(15) (20)(73)(74).

A further increase in wheel speed (in this case beyond 100 m/s) leads to a reduction in temperature. No research into this speed range has been carried out by other authors. The reasons for the change in direction of the temperature curve are discussed in detail in section 4.6.

The temperature was measured at different points of the contact zone to identify the position with the maximum thermal load. Figure 6.21 shows the contact-zone temperature curves at five different points measured at different specific removal rates.

At the relatively low specific removal rate $Q'_w = 10$ mm^3/(mm · s), the maximum temperature lies approximately in the middle of the contact

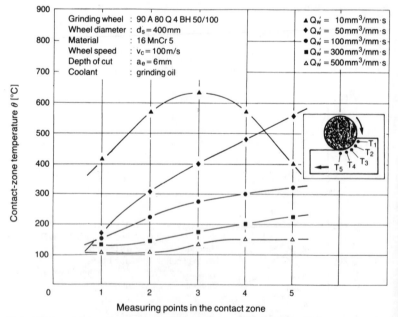

Fig. 6.21 **Contact-zone temperature at five different points in the contact zone measured at different specific removal rates (up-grinding)**

zone. This finding is in good agreement with the results of (20). A further increase in Q'_w shifts the temperature peak to measuring point 5, which is closest to the newly produced surface.

An increase in Q'_w (by means of v_w) brings about a marked reduction in temperature. For example, the maximum temperature at $Q'_w = 500$ mm^3/ (mm · s) amounts to only about 15 percent of the peak temperature at $Q'_w = 10$ mm^3/(mm · s). In down-grinding the temperatures at lower specific removal rates are higher than in up-grinding, but at higher Q'_w values the temperature is approximately the same in both grinding nodes (Fig. 6.30).

Figure 6.22 shows the peak temperatures at the 5 measuring points as they approach the contact zone. The grinding process was first interrupted at a distance of 10 mm in front of the zone containing the thermocouples, and the temperature was measured. Thereafter, grinding was continued to within 5 and 3 mm of measuring point 1, with the temperature being measured simultaneously. In the fourth and fifth test, the workpiece was ground to within 2 and 1 mm, respectively, of the contact

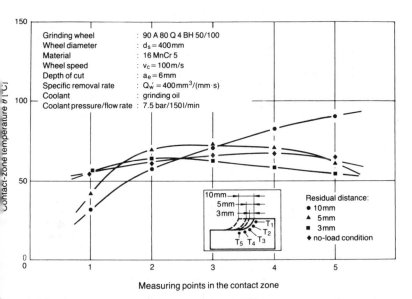

Grinding wheel	: 90 A 80 Q 4 BH 50/100
Wheel diameter	: $d_s = 400$ mm
Material	: 16 MnCr 5
Wheel speed	: $v_c = 100$ m/s
Depth of cut	: $a_e = 6$ mm
Specific removal rate	: $Q'_w = 400$ mm^3/(mm·s)
Coolant	: grinding oil
Coolant pressure/flow rate	: 7.5 bar/150 l/min

Residual distance:
● 10 mm
▲ 5 mm
■ 3 mm
♦ no-load condition

Measuring points in the contact zone

Fig. 6.22   Contact-zone temperature at different measuring points and with different grinding distances, i.e., different distances from the measuring points

zone. The last two results are shown by a single curve. In the test with a residual distance of 10 mm, the temperature is at its highest at measuring point 5, because the distance to this contact point is shorter and because at the higher specific removal rate $Q'_w = 400$ mm^3/(mm · s), the temperature at measuring point 5 is higher (Fig. 6.21).

In the subsequent tests under the same grinding conditions and with residual distances of 5, 2, and 1 mm, the temperature was lower, especially at point 5, despite the shorter distances to the measuring points (thermocouples). There are two reasons for this: in the first place, the total amount of energy converted is also lower because of the relatively smaller volume of stock removed, and, secondly, the grinding time is very short. Therefore only a small quantity of heat flows into the workpiece, and thus to the thermocouples.

After the material had been removed from the test piece up to the designated contact zone, further temperature measurements were carried out with the grinding wheel still rotating in the contact zone, but without any feed movement and therefore without any stock being removed. The grinding oil continued to be delivered to the contact zone and over the workpiece under the same conditions. Despite the relatively short time in

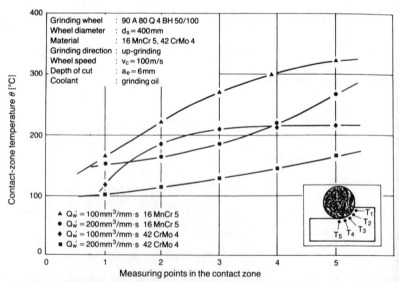

Fig. 6.23 Contact-zone temperature for two different materials at different specific removal rates

which the coolant was in the working area, the contact zone heated the coolant, supplied at room temperature, by about 50°C (Fig. 6.22).

By way of comparison, Fig. 6.23 shows the temperature curves in the contact zone for two different materials: normalised, low-alloy, heat-treated 42CrMo4 steel, and case-hardened 16MnCr5 steel in the untreated 'as supplied' condition. The materials were ground at two different specific removal rates, $Q'_w = 100$ and 200 mm^3/(mm · s). Again, the temperature was measured at five different points in the contact zone.

As with the previous results, the temperature falls as the specific removal rate increases, because of the faster workpiece speed. At both these rates the temperature is lower with the 42CrMo4 material than with 16MnCr5, because the lower-alloy steel is easier to grind.

6.6 RESIDUAL STRESSES AS A FUNCTION OF GRINDING PARAMETERS

The usual criteria for assessing the quality of a machined component are dimensional accuracy, geometrical accuracy and surface roughness. In many cases these characteristics are sufficient to ensure the quality of the workpiece. However, in the case of severely loaded or critical components they constitute only a description of the geometry and provide no information about the effect on the physical properties of the component, which are changed by the machining process.

Grinding – generally the final machining operation – alters the residual stresses set up in the material by a previous machining operation or hardening.

A knowledge of the state of residual stress can often provide a more precise insight into the physical characteristics of a component, and into the effects of machining, than temperature measurements or metallographic examinations. It is therefore essential to measure the residual stresses, particularly if it is intended to impart certain properties to the workpiece. As X-ray stress analysis can detect residual stresses in a thin surface zone (X-rays penetrate steel to a depth of 5–10 μm), and the depth of the stresses generated by grinding is of the same order of magnitude, this technique is particularly appropriate.

The internal stresses induced by grinding are determined by the following process variables (**97**)–(**103**):

– machine settings;
– grinding wheel: grit type, grit size, bond and bond hardness;

- coolant: type, quantity and pressure;
- dressing conditions;
- workpiece material;
- grinding method;
- wear behaviour of the grinding wheel.

In general it may be said that the process variables which produce higher contact-zone temperatures generate *tensile stresses* in the surface zone of the workpiece. These are usually undesirable, especially in components subject to dynamic loads. Machining parameters which give rise to higher mechanical forces set up compressive residual stresses (97)–(105). In the grinding process, heat and mechanical forces occur simultaneously, with the increase in grinding forces usually going hand in hand with a rise in temperature. Consequently, controlling the process so as to induce particular residual stresses is not a straightforward matter. Reducing the frictional forces generally leads to less heat being generated in the contact zone. The use of a suitable coolant, applied at an appropriate pressure and flow rate, lowers the grinding temperature, which can result in tensile stresses or possibly even compressive residual stresses. If an oil coolant is used, it does indeed reduce the grinding forces, but lacks the good cooling action of a coolant containing water (40)(46). In the case of materials where the cutting process involves high frictional force components, oil results in lower grinding temperatures by reducing the grinding forces, despite its limited cooling action. With materials where the frictional forces produced by grinding are lower, more effective cooling can be achieved by the superior heat dissipation properties of a coolant which contains water. Finer dressing conditions result in higher forces and an improved surface finish, but at the same time produce a higher grinding temperature, which shifts the stresses in the tensile direction (99).

CBN is generally known for its cooler cutting action and to produce stresses which tend to be compressive (7)(73)(101)–(103). A number of authors have noted tensile residual stresses with conventional aluminium oxide wheels, and compressive residual stresses with CBN wheels (73) (102)(103).

There are various reasons for the cooler cutting action of CBN wheels. The sharper grit and its superior thermal conductivity are the most important. The sharper grains give rise to lower friction and temperature, and, because of their higher thermal conductivity, the CBN grains convey large quantities of heat away from the contact zone

Another important reason for the cooler cutting action and reduced friction, and for shifting the residual stresses in the compressive direction, is the smaller rake angle with CBN grits. With Al_2O_3 grains the negative rake angle is 85–90 degrees, while with CBN grains it is 65–70 degrees (12).

The residual stresses in the grinding direction are generally more positive than those perpendicular to the grinding direction (7)(99)(101)(104) Fig. 6.24). As the thermal expansion is necessarily the same in all directions, the residual stresses caused by heat must also be the same in the grinding direction and perpendicular to it. This is not the case with residual compressive stresses.

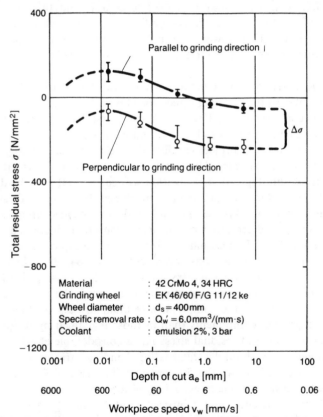

Fig. 6.24 Residual stress parallel and perpendicular to the grinding direction as a function of depth of cut (7)

Tönshoff and Brinksmeier (**101**) offer the following explanation. When grinding steel, the chip-forming process is characterised by an extremely high, all-round state of compressive stress in the area around the root of the chip. However, after the chip has been removed, some of this compressive stress is left behind, mainly at the sides of the individual groove produced while cutting. The workpiece material which was located in front of the advancing cutting edges is removed in the form of chips. The stresses present in this material are also eliminated or reduced. In other words, some of the stresses in the grinding direction are removed with the chips, resulting in higher (more positive) mechanically induced residual stresses perpendicular to the grinding direction.

The following point may be added to complement this explanation. The chip-forming temperature perpendicular to the grinding direction of each individual grit is lower than in the grinding direction, because the temperature in the area around the root of the chip falls more sharply at the sides.

Consequently, at low temperatures plastic deformation takes place perpendicular to the grinding direction. This, in turn, requires fairly high lateral mechanical forces. The greater transverse forces at the chip root cause the residual stresses to become compressive stresses.

In a joint research project conducted by the Production Engineering Department of Bremen University and the Bremen Institute for Materials Technology, an investigation was carried out into the dynamic strength behaviour of components machined by creep-feed grinding. Compared with workpieces ground with the reciprocating grinding method, creep-feed ground components respond to an increased depth of cut with a shift of the residual stresses into the compressive area in the various structural conditions investigated (**7**)(**104**)(**105**). Figure 6.24 shows the residual stress curve as a function of increasing depth of cut and the proportional reduction in workpiece speed. Despite the constant specific removal rate, the compressive residual stresses are higher in the creep-feed ground components than in those machined by reciprocating grinding. The state of residual stress has a considerable effect on the life and fatigue strength under alternating bending stresses of machined parts. Figure 6.25 shows the 50 percent failure-related fatigue strength under alternating bending stresses R_{BW} of the samples ground in the test in both the quenched and tempered, and hardened conditions. The increase in fatigue strength which accompanies increasing depth of cut is particularly marked in the hardened material, and is due to the higher compressive residual stresses.

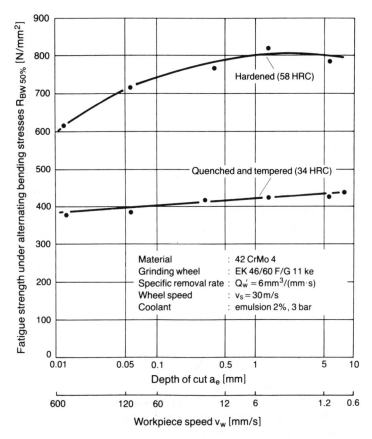

Fig. 6.25 Fatigue strength under alternating bending stresses as a function of depth of cut and pre-treatment of the workpiece (7)

Figure 6.26 shows the results of residual stress measurements on workpieces machined by high-efficiency deep grinding with CBN wheels. The grinding tests using a CBN wheel electroplated with B252 grit with worn abrasive peaks were carried out in the lower performance range for HEDG. Pure grinding oil was used as coolant.

While the number of tests carried out was too small to enable any generalised statement to be made which is valid in all cases, the results do, however, justify the following cautious and provisional conclusions.

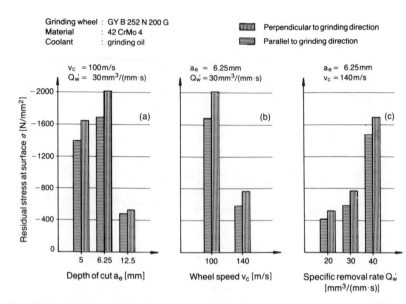

Fig. 6.26 High compressive residual stresses produced by HEDG with CBN wheels

- All the residual stresses are relatively high and lie in the compressive range.
- As the specific removal rate Q'_w increases, the compressive residual stresses also increase. This phenomenon can be explained by the increase in workpiece speed, resulting in lower grinding temperatures. In other words, because of the lower grinding temperature, the thermally determined tensile stresses are overcompensated by the mechanically determined compressive residual stresses (Fig. 6.26(c)).
- Increasing the peripheral speed of the wheel at relatively low specific removal rates in the lower performance range of high-efficiency deep grinding leads to higher temperatures, which shift the residual stresses in the tensile direction (Fig. 6.26(b)).

The very high compressive residual stresses in Fig. 6.26 cannot have been caused by any quenching during the grinding process, as they are greater than the residual stresses which can be obtained by quench-hardening. The exact causes need to be explained in greater detail by further investigations into the structural changes which occur, the relationship between hardness and depth, and the behaviour of the workpieces as finished components. The results also indicate that, if the grinding wheel and machining parameters are selected appropriately, some components do not need to be hardened, as the required physical surface-zone characteristics can be generated by the grinding operation itself.

6.7 INFLUENCE OF GRINDING DIRECTION ON RESULTS (DOWN- OR UP-GRINDING)

In conventional reciprocating and creep-feed grinding it is generally the case that down-grinding offers economic and technical advantages over up-grinding **(8)(20)**. Down-grinding brings improvements in surface finish, wear behaviour, and grinding ratios. The grinding forces are lower and the specific energy requirement is, therefore, correspondingly lower. Temperature measurements have produced approximately the same results with both methods **(9)(84)**.

The effect of the grinding mode on the machining result with HEDG was studied. Figure 6.27 shows two workpieces ground using the same machining parameters, one by up-grinding and the other by down-grinding. The thermal load was found to be greater on the workpieces machined by down-grinding. There are a number of reasons for this **(12) (34)(35)**.

Up-grinding

Down-grinding

Grinding wheel	: GY B252-N 200-G
Material	: 16 MnCr 5
Depth of cut	: $a_e = 5\,mm$
Specific removal rate	: $Q'_w = 50\,mm^3/(mm\cdot s)$
Wheel speed	: $v_c = 120\,m/s$
Coolant	: grinding oil

Fig. 6.27 Visual comparison between down- and up-grinding

(1) The coolant is more effective in up-grinding. This is clear from Fig. 6.28 (particularly at large depths of cut). This is because, in up-grinding, a sufficient quantity of fresh oil acts on the newly generated surface. In down-grinding, however, the coolant is introduced into the contact zone in a less advantageous manner and is heated as it passes through the contact zone before reaching the newly produced surface of the workpiece.

(2) In down-grinding, the hot chips, the heated coolant, and the heated surface of the grinding wheel come into direct contact with the newly produced surface of the workpiece.

(3) In down-grinding, particles of workpiece material lodged in the surface of the grinding wheel act spontaneously as sources of heat, even when they are discharged again on leaving the contact zone. In up-grinding, only those particles which have adhered firmly to the surface of the wheel act on the newly created workpiece surface.

This view of the thermally less favourable down-grinding mode is countered by the observation that the grinding forces in down-grinding are much lower than those in up-grinding (Fig. 6.29). This apparent contradiction is explained by the fact that, in down-grinding, the chip-forming process with each engaged cutting edge starts with the maximum chip cross-section, which decreases to zero as the process of chip formation proceeds. In up-grinding, the sequence is exactly the reverse. In down-grinding, therefore, chip-formation starts at once, whereas in up-grinding an intensive frictional and deformation phase

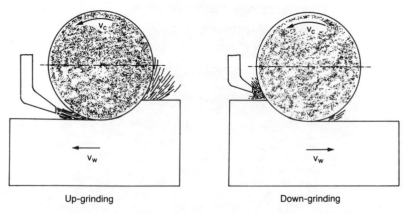

    Up-grinding   Down-grinding

Fig. 6.28 Coolant delivery in up- and down-grinding

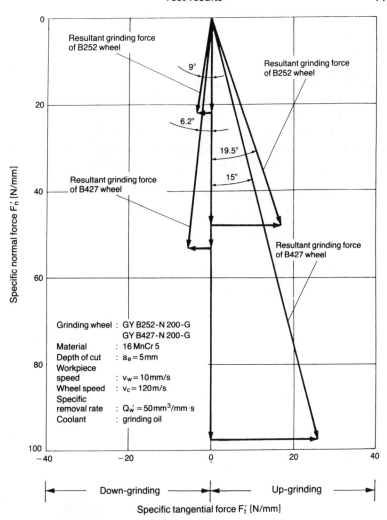

Fig. 6.29 Comparison of grinding force components in up- and down grinding with different CBN grits

(along with accompanying high chip-forming forces) has to be completed before the chip starts to form (**12**).

The difference between the grinding forces involved in down- and up-grinding diminishes as the specific removal rate increases, i.e., at very high removal rates the grit engages the workpiece fairly rapidly, even in up-grinding, and the frictional and deformation phases are eliminated.

Figure 6.28 shows that in up-grinding, because the coolant is not obstructed, the areas of the workpiece heated by machining are subjected to an intense cooling effect. The whole process may be compared with a clearly defined quenching action (9).

Another series of tests was carried out to investigate the influence of down- and up-grinding on the development of temperature in the contact zone produced by machining at different specific removal rates (Fig. 6.30). The temperature was measured at five different points within

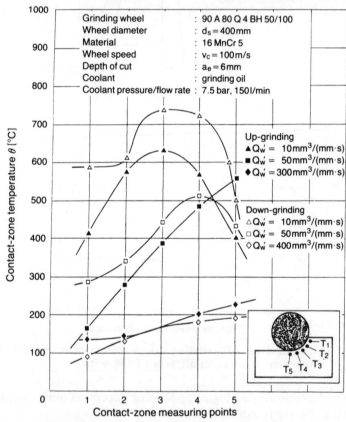

Fig. 6.30 Temperature in the contact zone and at the newly created surface (up- and down-grinding)

the contact zone. Thermocouple 1 (point 1) was located at the uppermost position in the contact zone. Point 5, marking the fifth measuring point, was situated at the lowest position in the contact zone (the newly created surface). Figure 6.30 shows that the contact-zone temperature rises as Q'_w decreases (this does not apply at very low Q'_w values).

When Q'_w is low, the contact-zone temperatures in down-grinding are higher than those occurring in up-grinding.

The point in the contact zone where the temperature is at its highest varies as a function of Q'_w. Machining at a low specific removal rate Q'_w of 10 mm^3/(mm · s), as is normal in creep-feed grinding, the maximum temperature is found in the middle of the contact zone. These results confirmed the findings of other authors **(15)(20)(84)**. In HEDG, on the other hand, the maximum temperature, though lower, occurs at the newly created surface. The temperatures are approximately the same both in up- and down-grinding. Wider grinding wheels are generally liable to produce higher temperatures, especially in down-grinding.

Figure 6.27 shows a typical example of this. Burn marks are clearly visible on a workpiece machined by down-grinding with a CBN wheel ($b_s = 8$ mm).

From these results, the following recommendations can be made with regard to the choice between up- and down-grinding.

- For medium specific removal rates ($Q'_w = 20$–70 mm^3/(mm · s)), up-grinding should be chosen because the workpiece sustains less thermal damage.
- For low specific removal rates ($Q'_w < 20$ mm^3/(mm · s)), which correspond more or less to those employed in conventional grinding, and for high specific removal rates ($Q'_w > 70$ mm^3/(mm · s)), it is generally better practice to operate in the down-grinding mode, as the specific grinding forces are lower and the wheel wears more slowly.

6.8 INFLUENCE OF GRIT SIZE ON THE GRINDING RESULT

A coarser grit generally results in greater chip thicknesses, leading in turn to lower grinding forces **(14)**. Some of the results obtained in the tests varied considerably.

Tests were initially carried out with two electroplated CBN wheels (400 mm diameter, 8 mm wide abrasive layer), one with a medium and the other with a coarse grit (B252 and B427). These were followed by tests with two narrower CBN wheels (2 mm wide abrasive layer) containing B252 and B151 grit. The bond type and wheel diameter remained unchanged. The results of the tests with the first two wheels may be summarised as below.

Because of the shallow rectangular shape of the ground grooves, there were no axial force components. Figure 6.31 shows the force components measured at a depth of cut a_e of 10.0 mm for the two different CBN grits in relation to wheel speed v_c. The following conclusions can be drawn.

(a) As wheel speed v_c increases, the normal and tangential force components steadily decrease.

(b) The tangential force component is always lower than the normal force component.

(c) The grinding forces for the coarse grit (B427) are appreciably higher than those for the finer grit (B252).

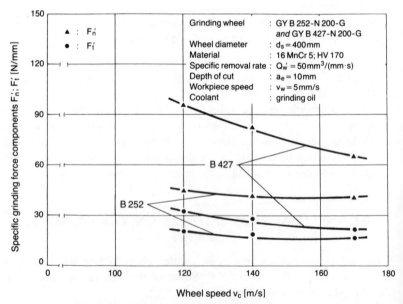

Fig. 6.31 Specific grinding force components as a function of wheel speed for different grit sizes

This latter observation is unexpected and is the first clear evidence of a phenomenon which could result in the rejection of a notion which hitherto has been widely accepted **(12)(34)(35)**.

Until now it has been assumed that wheels containing a fine grit will always give rise to higher cutting forces, because of the greater number of cutting edges (under otherwise constant process parameters), as the frictional force component increases if the chip cross-section is small. However, this assumption holds only if the wear process on the abrasive grits being compared is identical and if the chip-formation process takes place under quite specific kinematic–geometrical conditions and specific stress conditions.

A detailed investigation of these complex relationships must be deferred for later research.

Figure 6.32 compares the grinding forces relative to specific removal rate for two different grit sizes. In this graph the finer CBN grit (B151) is characterised by higher grinding forces. The higher grinding forces

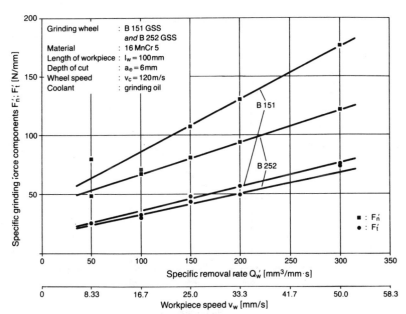

Fig. 6.32 Specific grinding force components for fine grit (B151) and medium grit (B252) at different specific removal rates

encountered with the finer CBN grit, as compared with medium B252 grit, are in line with expectations.

At a fairly high specific removal rate ($Q'_w > 300$ mm³/(mm · s)), the difference in grinding forces with B252 and B151 grits is less marked (Figs 6.7 and 6.8).

The following comments can be made with regard to the higher grinding forces observed with coarser CBN grits.

The larger average chip cross-sections produced by a coarse grit result in a thicker contact layer. When this layer exceeds a certain thickness, more time and energy are needed to attain equilibrium temperature in the lower contact layer, and this in turn generates higher grinding forces (section 4.6).

The manner in which grinding forces tend to increase relative to grit size in CBN wheels is shown in Fig. 6.33.

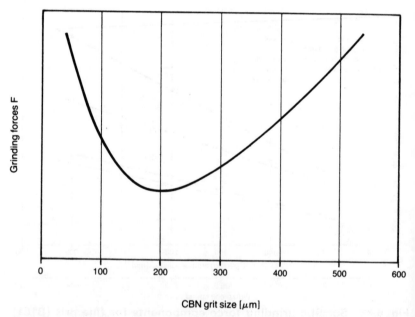

CBN grit size [μm]

Fig. 6.33 Grinding forces relative to grit size for CBN grinding wheels

Overall, it can be said that the forces occurring in high-efficiency deep grinding are at their lowest when medium grit sizes are used. CBN grits finer or coarser than about B200 give rise to higher grinding forces, resulting in greater tool wear, greater geometrical errors and, usually, higher grinding temperatures. A medium grit should therefore be chosen in preference to a coarser or finer grit, provided that this meets the surface finish requirements of the machining task.

7

Summary

High-efficiency deep grinding (HEDG) is a further development of the creep-feed grinding technique involving increased peripheral wheel speeds. Its main feature lies in the combination of high removal rates and high wheel speeds, and is therefore characterised by large depths of cut, high workpiece speeds, and high peripheral wheel speeds. The technique enables grooves and profiles several millimetres deep to be produced in a single pass, imparting excellent machining quality, roughness characteristics and structural integrity. Furthermore, the removal rates which can be achieved are far greater than those obtainable by turning or milling.

The object of this study has been to analyse the technical and mechanical aspects of high-efficiency deep grinding, and to verify these on the basis of extensive practical tests. As far as the requirements in terms of the grinding machine and the wheel itself are concerned, it has been established that conventional machines are inadequate. Rigid machines capable of high spindle outputs and speeds, together with suitable coolant delivery systems and appropriate grinding wheels, are essential if full benefit is to be derived from the technique.

The extremely high removal rates achieved in HEDG require a considerable amount of energy to be converted in the contact zone between the workpiece and the grinding wheel. This process energy is mainly converted into heat, only a small proportion of which must flow into the workpiece. A suitable coolant must be used and delivered in a form appropriate to the process in order to remove the process heat away from the grinding zone. At the same time, the particles of workpiece material adhering to the surface of the wheel must be dislodged by coolant delivered at a high pressure. The design of the coolant delivery system (the shape of the nozzle, pressure, flow-rate) has been described in detail.

Another important factor in HEDG is the conditioning of the wheel by dressing, since its working profile and the sharpness of the wheel must be maintained at all times. This was achieved with a so-called swing-step dressing apparatus fitted with force measuring sensors and with suitable drives and controls to optimise the process. The mechanical principles and kinematic requirements underlying this technique were described and compared with conventional dressing methods.

When dressing CBN wheels, the cutting ability of the abrasive grains must be ensured by a subsequent sharpening process. In this context the influence of the shape of the chip space was discussed in quantitative terms. It was demonstrated that the optimum average chip space geometry was created by the process of chip formation, rather than by sharpening. However, sharpening is necessary to initially open up the surface of the grinding wheel.

Despite extremely high removal rates of up to 3000 mm^3/(mm · s), even HEDG is subject to limits which are imposed by the chip space available per unit of time. The analytical model which was developed enables the maximum specific removal rate to be calculated as a function of the process parameters. The results of the calculations were confirmed by practical tests.

Besides conventional bakelite and vitrified bond grinding wheels containing Al$_2$O$_3$ grit, metal bond CBN wheels have proved particularly suitable for high-efficiency deep grinding. HEDG tests were performed to determine grinding forces, wheel wear, and machining results. Conventional bakelite bond aluminium oxide wheels are also suitable for high-efficiency deep grinding. They wear relatively slowly (a high *G* ratio), can be dressed easily and show little tendency to become loaded. If conventional wheels are to be used in HEDG, it is extremely important that they must have sufficient resistance to bursting, as they must be able to operate at peripheral wheel speeds v_c in excess of 80 m/s.

The main section of the book has addressed the technological and mechanical features of HEDG and how they enable very high material removal rates to be achieved. Attention was focused on discussing hitherto unresolved questions regarding grinding forces, process temperatures, and the effect of the type of abrasive used in HEDG. A new approach to clarifying these matters was introduced with the contact layer theory, which postulates that between the wheel and the workpiece there is a layer of interaction which is relevant to the process, the thickness of which corresponds to the average chip thickness.

As higher process temperatures are generally to be expected at high removal rates and high levels of energy conversion, the temperatures occurring in the newly formed surface zone of the workpiece and in the contact zone were measured with micro-thermocouples. This served to disprove the generally held view that the surface-zone temperature increases continuously as the wheel speed increases. Tests showed that, both with conventional and CBN wheels, the surface-zone temperature of the workpiece increases only up to wheel speeds of around 100 m/s. Once the peripheral speed exceeds this limit, the surface-layer temperature passes through a maximum and then falls. This observation is of great importance for the development of high-efficiency deep grinding. It is surprising, but can be explained by the contact layer theory.

The low process temperature in HEDG is also important with regard to the way in which residual stresses are set up in the surface zone of the workpiece. The low temperature means that the picture is dominated by compressive residual stresses generated by plastic deformation at temperatures which are low or prevail for brief periods. Residual stress measurements taken at various points on the surface of steel structural elements ground with CBN wheels at very high removal rates produce remarkable results. Extremely high compressive residual stresses are found, which in some cases are substantially greater than the strength of the material. This study has been the first to establish such high compressive residual stresses in a low-alloy steel. This surprising result could indicate that in certain cases, if grinding is carried out in a controlled manner, it could take the place of thermal hardening.

The final section describes the results of grinding tests in terms of the main evaluation criteria (grinding force, performance, surface finish, wear, grinding ratio, temperatures, and residual stresses) relative to the main process parameters. The effect of grinding direction (up- and down-grinding) and grit size is also discussed.

The findings of this research should enable machine and tool manufacturers, as well as end-users, to plan and implement suitable high-efficiency deep grinding processes and to develop equipment best suited to reap the benefits of this technology.

References

(1) M. E. MERCHANT, Forecast of the future of production engineering, *Annals CIRP*, 1971, **19**.

(2) G. GEISWEID and W. GÄRTNER, Tief- und Pendelschleifen, Temperaturen und Energiebedarf, *Industrie Diamanten Rundschau*, 1978, **12**(2), 105–108.

(3) G. R. SHAFTO, *Creep-feed grinding – an investigation of surface grinding with high depth of cut and low feed rates*, Phd Thesis, University of Bristol, UK, 1975.

(4) G. WERNER, Realisierung niedriger Werkstückoberflächentemperaturen durch den Einsatz des Tiefschleifens, *Trenn-Kompendium*, (ETF-Verlag, Bergisch-Gladbach, Germany), 1983, Vol. 2, pp. 448–468.

(5) G. WERNER, Application and technical fundamentals of deep and creep-feed grinding; SME Technical Paper MR 79-319, 1979.

(6) G. WERNER and E. MINKE, Technologische Merkmale des Tiefschleifens – Erhöhte Schnittkräfte und reduzierte Werkstücktemperaturen, *tz f. Metallbearb.* (Part 1), 1981, **75**(3), 11–15; (Part 2), 1981, **75**(5), 44–48.

(7) G. WERNER and T. TAWAKOLI, Der Druckeigenspannungszustand tiefgeschliffener Stahlbauteile, *tz f. Metallbearb.*, 1986, **80**(6), 39–46.

(8) W. KÖNIG, *Fertigungsverfahren* (vol. 2, Schleifen, Honen, Läppen) (VDI-Verlag GmbH, Düsseldorf, Germany) 1980.

(9) R. LOWIN, *Schleiftemperaturen und ihre Auswirkungen im Werkstück*, Dissertation, RWTH Aachen, Germany, 1980.

(10) H. BRANDIN, *Pendelschleifen und Tiefschleifen – Vergleichende Untersuchungen beim Schleifen von Rechteckprofilen*, Dissertation, TU Braunschweig, Germany, 1978.

(11) K. GÜHRING, *'Hochleistungs-Schleifen', eine Methode zur Leistungssteigerung der Schleifverfahren durch hohe Schnittgeschwindigkeit*, Dissertation, RWTH Aachen, Germany, 1967.

(12) G. WERNER and T. TAWAKOLI, Fortschritte beim HEDG-Verfahren mit CBN-Schleifscheiben, *Industrie Diamanten Rundschau*, 1988, **22**(1), 17–24.

(13) K. GÜHRING, Anwendung hoher Schnittgeschwindigkeiten beim Schleifen mit bakelitisch gebundenen konventionellen Schleifscheiben, TAE-seminar 'High-speed machining', Ostfildern, Germany, 1985.

(14) G. KASSEN, *Beschreibung der elementaren Kinematik des Schleifvorganges*, Dissertation, RWTH Aachen, Germany, 1969.

(15) G. WERNER, *Kinematik und Mechanik des Schleifprozesses*, Dissertation, RWTH Aachen, Germany, 1971.

(16) G. WERNER, Hohe Abtragsleistungen durch Bearbeitungsgüten durch moderne Schleifverfahren, *Automobil-Industrie*, 1988, **33**(2), 169–177.

(17) G. WERNER, Untersuchung zur Optimierung des Kühlschmierstoffeinsatzes an Hochleistungstiefschleifmaschinen für Flachschleifoperationen, Interim report on AIF research project No. FE 3060, January 1986.

(18) U. UHLIG, W. REDEKER and R. BLEICH, Profilschleifen mit kontinuierlichem Abrichten, *Werkstatttechnik*, 1982, **72**(6), 313–317.

(19) W. KÖNIG and A. ARCISZEWSKI, Schleifen mit kontinuierlichem Abrichten von schwer zerspanbaren Stählen; *Industrie-Anzeiger*, 1987, **110**(61/62), 26–29.

(20) C. ANDREW, T. D. HOWES and T. R. A. PEARCE, *Creep-feed grinding*; (Holt, Rinehart and Winston, London, New York, Sydney, Toronto), 1985.

(21) W. KÖNIG, J. FROMLOWITZ, B. STUCKENHOLZ and K. YEGENOGLU, Hochleistungsschleifwerkzeuge, *VDI-Z*, 1985, **127**(21), 845–848.

(22) H. R. MEYER, F. KLOCKE and J. SAUREN, An overview of the technology of CBN grinding, 2nd Intl. Grinding Conference (SME), 1986, Philadelphia, PA, USA.

(23) K. YEGENOGLU, *Berechnung von Topographiekenngrössen zur Auslegung von CBN-Schleifprozessen*, Dissertation, RWTH Aachen, Germany, 1986.

(24) B. A. COOLEY and H. WAPLER, Grundlegende Zusammenhänge beim Einsatz von abrasiven Bornitrid, Diamant-Information M31 (De Beers Industrie-Diamanten (Deutschland) GmbH, Düsseldorf, Germany).

(25) H. HELLETSBERGER and N. BOGNER, CBN versus conventional abrasives; 2nd Intl. Grinding Conference (SME), 1986, Philadelphia, PA USA, (SME Technical Report MRR 86-03).

(26) R. P. LINDSAY, Laboratory investigations in support of CBN production grinding; Norton, Worcester, MA, USA.

(27) H. D. DODD and D. V. KUMAR, Technological fundamentals of CBN bevel gear finish grinding, Conference of Superabrasives, 1985, Chicago, IL, USA (SME, MR 85-273).

(28) H.-R. MEYER, F. KLOCKE and J. SAUREN, Hochgeschwindigkeitsschleifen mit CBN-Schleifscheiben, *5th Internationales Braunschweiger Feinbearbeitungskolloquium*, 1987.

(29) W. KÖNIG, K. STEFFENS, J. FROMLOWITZ, K. YEGENOGLU, and B. STUCKENHOLZ, Leistungspotentiale beim Schleifen, *Industrie-Anzeiger*, 1985, **107**(72), 177–179.

(30) H. STREITMÜLLER, *Kinematische Grundlagen für die praktische Anwendung des spitzenlosen Hochleistungsschleifens*, Dissertation, RWTH Aachen, Germany, 1971.

(31) H. K. TÖNSHOFF and P. DENNIS, Hochleistungsbandschleifen – eine Alternative zum Fräsen, *Industrie-Anzeiger*, 1987, **109**(67), 38–39.

(32) W. KÖNIG, H. K. TÖNSHOFF, J. FROMLOWITZ and P. DENNIS, Belt grinding; *Annals CIRP*, 1986, **35**(2), 487–494.

(33) G. WERNER, T. TAWAKOLI and E. MINKE, Schmal und tief – Hochgeschwindigkeitsschleifen enger Nuten mit kubischem Bornitrid, *Maschinenmarkt*, 1988, **94**(9), 20–25.

(34) G. WERNER and T. TAWAKOLI, Advances in high-efficiency deep grinding (HEDG), SME Technical Paper, MR 88-588, 1988.

(35) T. TAWAKOLI, 180 m/s mit galvanisch gebundenen CBN-Schleifscheiben, presented at *HSG-Technologie und CBN-Schleifwerkzeuge – eine neue Dimension in der Metallbearbeitung*, (Diamant Boart and Gühring Automation, Stetten-Frohnstetten), 1988.

(36) G. WERNER and T. TAWAKOLI, Hochleistungsschleifen von engen Schlitzen mit CBN-Schleifscheiben, *Industrie Diamanten Rundschau*, 1988, **22**(2), 91–95.

(37) G. WERNER and T. TAWAKOLI, High-efficiency deep grinding with CBN; *Industrial Diamond Review*, 1988, **48**(3).

(38) H. FRANK, Sicherheit von Schleifkörpern, *Industrie-Anzeiger*, 1985, **107**(1/2), 28–31.

(39) H. INOUE, H. NAGUCHI, Y. TAKAHASHI and S. SUZUKI, *Research on high-efficiency grinding, new development in grinding* (Edited by M. C. Shaw) (Carnegie Press, Carnegie-Mellon University, Pittsburgh, PA, USA) 1972.

(40) R. VITS, *Technologische Aspekte der Kühlschmierung beim Schleifen*, Dissertation, RWTH Aachen, Germany, 1985.

(41) E. TRAUTVETTER, Kühlschmierstoffe für die spanende Metallbearbeitung, *Trenn-Kompendium*, (ETF-Verlag, Bergisch-Gladbach, Germany), 1983, Vol. 2.

(42) T. TAWAKOLI, Anforderungen an Kühlschmierstoff-Anlagen und Zufuhrsysteme

beim Hochleistungsschleifen, Seminar on 'Coolants in metalworking' organized by the Deutsches Industrieforum für Technologie, 1992, Düsseldorf, Germany.

(43) W. KÖNIG and R. VITS, Kühlschmierstoffzuführung beim Aussenrundschleifen, *Jahrbuch Schleifen, Honen, Läppen und Polieren*, 54th edition, (Vulkan-Verlag, Essen, Germany), 1987.

(44) T. GRABNER, Leistungssteigerung bei keramisch gebundenen CBN-Schleifscheiben durch Hochdruckkühlschmierung, *Industrie-Anzeiger*, 1987, **109**(43/44), 61–62.

(45) W. KÖNIG, F. U. MEIS, L. NEDER, P. SARTORI, G. HOLTUS and H. JOHANSSEN, *Schadstoffe beim Schleifvorgang*, (Bundesanstalt für Arbeitsschutz, Dortmund, Germany), Report No. 427, 1985.

(46) G. WERNER and H. LAUER-SCHMALTZ, *Advanced application of coolants and prevention of wheel loading in grinding metalworking lubrication* (ASME, New York), pp. 225–232, 1980.

(47) H. W. OTT, Kühlschmieren – Voraussetzungen für kostengünstiges Schleifen und Abrichten, presented at the VDI seminar 'Schleifen als qualitätsbestimmende Endbearbeitung', 1982, Stuttgart, Germany.

(48) E. SALJÉ and J. RIEFENSTAHL, Kühlmittelzufuhr durch die Schleifscheibe beim Innenrundschleifen, *Industrie-Anzeiger*, 1982, **104**(53), 39–40.

(49) H. LAUER-SCHMALTZ, *Zusetzung von Schleifscheiben*, Dissertation, RWTH Aachen, Germany, 1979.

(50) F. SPERLING, *Grundlegende Untersuchungen beim Flachschleifen mit hohen Schleifscheibenumfangsgeschwindigkeiten und Zerspanleistungen*, Dissertation, RWTH Aachen, Germany, 1970.

(51) K. ELBEL, Acrylharzgebundene Schleifkörper steigern Produktivität und Genauigkeit (Acrylic resin bonded grinding wheels increase productivity and accuracy); *Werkstatt und Betrieb*, (1987), **120**(9), 717–720.

(52) G. WARNECKE, F.-J. GRÜN and K. ELBEL, Richtig Schmieren, PMMA-gebundene Scheibe mit Metallseife arbeitet mit kühlem Schliff und erzielt gute Oberflächenwerte, *Maschinenmarkt*, 1987, **93**(21), 26–32.

(53) H. K. TÖNSHOFF and W. GÄRTNER, Übersicht über die Möglichkeiten und Grenzen des Profilierens und Abrichtens von Diamant- und Bornitrid-Schleifscheiben, *Industrie Diamanten Rundschau*, 1981, **15**(4), 212–218.

(54) E. MINKE, *Grundlagen der Verschleissausbildung an nicht-rotierenden Abrichtschneiden zum Einsatz an konventionellen Schleifwerkzeugen*, Dissertation, Bremen University, Germany, 1988

(55) J. MESSER, *Abrichten konventioneller Schleifscheiben mit stehenden Werkzeugen*, Dissertation, RWTH Aachen, Germany, 1983.

(56) H. R. MEYER, Über das Abrichten von Diamant- und CBN-Schleifwerkzeugen, *Jahrbuch Schleifen, Honen, Läppen und Polieren*, 50th edition, (Vulkan-Verlag, Essen, Germany), pp. 312–331, 1981.

(57) W. KÖNIG and B. STUCKENHOLZ, Touch dressing of CBN wheels (TDC), Bessere Oberflächen durch gezieltes Abrichten, *Industrie-Anzeiger*, 1986, **108**(63/64), 27–28.

(58) H.-R. MEYER and J. SAUREN, Keramisch gebundenes CBN beim Innenrundschleifen, *Industrie-Anzeiger*, 1987, **109**(61/62), 18–25.

(59) G. WARNECKE and F. J. GRÜN, Abrichten kunstharzgebundener CBN-Schleifscheiben, *VDI-Z*, 1987, **129**(3), 80–85.

(60) G. WARNECKE, F.-J. GRÜN and W. GEIS-DRESCHER, Anwendung von PKD-Schneidplatten in Abrichtrollen, *Industrie Diamanten Rundschau*, 1988, **22**(1), 25–30.

(61) B. STUCKENHOLZ, *Das Abrichten von CBN-Schleifscheiben mit kleinen Abrichtzustellungen*, Dissertation, RWTH Aachen, Germany, 1988.

(62) H. K. TÖNSHOFF and T. GRABNER, Piezo-Abrichten – ein Abrichtsystem für CBN-Schleifscheiben, *VDI-Z*, 1988, **130**(6), 66–68.

(63) H. SCHLEICH, *Schärfen von Bornitrid-Schleifscheiben*, Dissertation, RWTH Aachen, Germany, 1982.

(64) W. KÖNIG and H. SCHLEICH, Abrichten und Schärfen von CBN-Schleifscheiben, *Industrie Diamanten Rundschau*, 1983, **17**(2), 68–78.

(65) R. HOLZ and J. SAUREN, *Schleiftechnisches Handbuch*, (Ernst Winter & Sohn, Germany), 1986.

(66) E. SALJÉ, H. MÖHLEN and U. MERTENS, Strahlschärfen von CBN- und Diamant-Schleifscheiben, *Industrie Diamanten Rundschau*, 1987, **21**(3), 180–183.

(67) G. STÄHLI, Die hochenergetische Kurzzeit-Oberflächenhärtung von Stahl mittels Elektronenstrahl-, Hochfrequenz- und Reibimpulsen, *Härterei-Technische Mitteilungen*, 1974, **29**(2), 55–57.

(68) J. F. KAHLES and D. EYLON, Machining of titanium alloys *J. Metals*, 1985, **37**(4), 27–35.

(69) W. HÖNSCHEID, *Abgrenzung werkstoffgerechter Schleifbedingungen für die Titanlegierung TiAl6V4 (Definition of grinding parameters suitable for titanium alloy TiAl6V4)*; Dissertation, RWTH Aachen, Germany, 1975.

(70) J. TRIEMEL, *Untersuchung zum Stirnschleifen von Schnellarbeitsstählen mit Bornitridwerkzeugen (Investigation into the face grinding of high-speed steels with cubic boron nitride wheels)*; Dissertation, TU Hanover, Germany, 1975.

(71) DUBBEL, *Mechanical engineering manual*, 14th edition (Springer-Verlag, Berlin, Heidelberg, New York), 1981.

(72) O. DAUDE, *Untersuchung des Schleifprozesses – Zusammenhang Schleifscheibe, Bearbeitungsbedingungen und Arbeitsergebnis*, Dissertation, RWTH Aachen, Germany, 1966.

(73) H. Z. CHOI, *Beitrag zur Ursachenanalyse der Randzonenbeeinflussung beim Schleifen*, Dissertation, Hanover University, Germany, 1986.

(74) H. E. GROF, *Beitrag zur Klärung des Trennvorgangs beim Schleifen von Metallen*, Dissertation, TU Munich, Germany, 1977.

(75) K. STEFFENS, *Thermomechanik des Schleifens*, Dissertation, RWTH Aachen, Germany, 1983.

(76) K. TAKAZAWA, Effects of grinding variables on surface structure of hardened steel; *Bull. Japan Soc. Precision Engng*, 1966, **2**(1), 14–21.

(77) M. KAISER, Thermoelektrische Erfassung von Schleifkörnern und Bindungsbrücken zur Untersuchung des Schleifprozesses, *Industrie-Anzeiger*, 1975, **97**(28), 549–550.

(78) D. G. LEE, *An experimental study of thermal aspects of grinding*, Dissertation, University of Cincinnati, OH, USA, 1971.

(79) S. MALKIN, *The attritious and fracture wear of grinding wheels*, Dissertation, Massachusetts Institute of Technology, Boston, MA, USA, 1968.

(80) G. WERNER, Schleifscheiben-Spezifikation und Werkstückstoff als bestimmende Merkmale für anwendbare Schnittgeschwindigkeit und Zeitspanungsvolumina, *Jahrbuch Schleifen, Honen, Läppen und Polieren*, 49th edition, (Vulkan-Verlag, Essen, Germany), 1979.

(81) W. J. SAUER, *Thermal aspects of grinding*, Dissertation, Carnegie Mellon University, Pittsburgh, PA, USA, 1971.

(82) S. A. POPOV and V. M. DAVYDOV, Contactless method of temperature measurement in grinding, *Russian Engng J.*, 1969, **49**(1), 74–77.

(83) G. WERNER, *Konzept und technologische Grundlagen zur adaptiven Prozeßoptimierung des Aussenrundschleifens*, Inaugural dissertation, RWTH Aachen, Germany, 1973.

(84) H. H. DAMLOS, *Prozeßablauf und Schleifergebnisse beim Tief- und Pendelschleifen von Profilen*, Dissertation, TU Braunschweig, Germany, 1984.

(85) J. GÖRNE, *Simulationsmodell zur Prozeßauslegung beim Schrägeinstechschleifen*, Dissertation, RWTH Aachen, Germany, 1986.

(86) V. A. SIPAILOV, Calculating grinding temperatures, *Russian Engng J.*, 1966, **46**(8), 78–79.

(87) F. RICHTER, *Die wichtigsten physikalischen Eigenschaften von 52 Eisenwerkstoffen*, (Verlag Stahleisen mbH, Düsseldorf, Germany), 1973.

(88) G. WERNER and MINKE, E. Weiterentwicklung analytischer und praktischer Methoden zur Bestimmung der Oberflächentopographie von Schleifscheiben (Further development of analytical and practical methods for determining the surface topography of grinding wheels), Final report of FNK research project No. 471, Bremen University, Germany, 1985.

(89) *Temperaturmessungen mit Miniatur-Mantel-Thermoelementen*, Philips Elektronic Industrie GmbH, Kassel, Germany.

(90) *Thermodrähte, Thermopaare*, Heraeus company product literature, Germany, 1984.

(91) M. KURREIN, Die Messung der Schleifkraft, *Werkstattstechnik*, 1927, 20, 585–594.

(92) W. ERNST, *Erhöhte Schnittgeschwindigkeit beim Aussenrund-Einstechschleifen und ihr Einfluss auf das Schleifergebnics und die Wirtschaftlichkeit*. Dissertation, RWTH Aachen, Germany, 1964.

(93) J. F. PRINS, Wechselwirkung zwischen Diamanten und Werkstückstoffen in Einkornversuchen, *Diamant-Information M24*, (De Beers Industrie-Diamanten (Deutschland) GmbH, Düsseldorf, Germany).

(94) G. WERNER and E. MINKE, Die Bezugs-Grundrauheit, funktionale Ableitung und praktische Anwendung, *VDI-Z*, 1986, **128**(11), 429–438.

(95) T. TAWAKOLI and S. J. TAVAKKOLI, High-efficiency deep grinding (HEDG) with shaved CBN wheels, SME technical paper MR90-506, 1990.

(96) T. TAWAKOLI, H. MAFTOON and S. J. TAVAKKOLI, Hochleistungsschleifen mit touchierten CBN-Schleifscheiben, *Industrie Diamanten Rundschau*, 1991, **25**(2), 108–113.

(97) E. MACHERAUCH, H. WOHLFAHRT and U. WOLFSTIEG, Zur zweckmässigen Definition von Eigenspannungen, *Härterei-Technische Mitteilungen*, 1973, **28**(3), 201–211.

(98) P. G. ALTHAUS, Werkstückeigenspannungen beim Einsatz von CBN- und Korundschleifscheiben zum Innenschleifen, *Industrie Diamanten Rundschau*, 1983, **17**(4), 184–190.

(99) E. BRINKSMEIER, *Randzonenanalyse geschliffener Werkstücke*, Dissertation, Hanover University, Germany, 1982.

(100) W. KÖNIG and R. LOWIN, *Ermittlung des Eigenspannungszustandes in der Randzone geschliffener Werkstücke und Bestimmung seiner Auswirkung auf das Funktionsverhalten*, Research reports of the Land North-Rhine Westphalia, No. 2886, (Westdeutscher Verlag, Opladen, Germany), 1979.

(101) H. K. TÖNSHOFF and E. BRINKSMEIER, Eigenspannungen durch Schleifen – wesentliche Einflussgrössen des Prozesses, *Eigenspannungen, Entstehung – Messung – Bewertung*, (E. MACHERAUCH and V. HAUK, Deutsche Gesellschaft für Metallkunde e.V., Germany), Vol. 2, pp. 251–270, 1983.

(102) R. AERENS and J. PETERS, Inducing residual compressive stresses in cylindrical face grinding by using CBN wheels, *Jahrbuch Schleifen, Honen, Läppen und Polieren*, 54th edition, (Vulkan-Verlag, Essen, Germany), 1987.

(103) G. A. JOHNSON, Beneficial compressive residual stress resulting from CBN grinding; SME Second International Grinding Conference, 1986, Philadelphia, PA, USA.

(104) G. WERNER, T. TAWAKOLI, P. MAYR and V. SPEICHER, Compressive

residual stresses in creep-feed ground work surfaces, SME Technical Paper, MR 86-635, 1986.

(105) P. MAYR and V. SPEICHER, Schwingfestigkeitsverhalten tiefgeschliffener Bauteile, Final report of DFG project MA 937/3–3, Germany, 1987.

Index